# UNBUILDING

# UNBUILDING

## SALVAGING THE ARCHITECTURAL TREASURES OF UNWANTED HOUSES

BOB FALK | BRAD GUY

UNBUILDING

The Taunton Press

Photographs ©Sally Kamprath: pp. 6–7, 15 (top), 21, 61, 68 (bottom), 69 (bottom), 70, 73 (top), 75 (bottom), 76 (top left), 77 (all except top right, bottom right), 78 (top left), 80 (top left), 82 (top left and right), 86 (bottom), 192; Brad Guy: pp. vi, vii–viii, 5, 9 (bottom), 10–11, 13, 30 (bottom), 31, 35, 38, 42 (bottom left), 48, 49–53, 55–60, 66–67, 68 (top), 71, 72 (bottom), 75 (top right), 76 (bottom), 77 (bottom right), 78 (center, bottom), 79 (all except top left), 81 (bottom), 82 (center, bottom), 84 (top left, center), 85, 86 (center), 87 (top), 88 (bottom), 91 (top), 93, 96 (center), 100, 106 (bottom), 107 (top), 110, 112 (top), 113 (top), 115–116, 118, 121, 122 (bottom), 123 (center, bottom), 124 (top left, top right), 126, 127 (top, bottom right), 128 (bottom left, center), 129 (top left, right), 130 (bottom), 132, 134, 135 (right), 141, 145 (center), 181 (right), 182–183, 184 (right), 189, 190, 194 (bottom), 197, 199, 200 (top, center), 201 (bottom), 203, 212–214, 218 (top left), 219, 221 (all except top left), 224 (all except top right), 225, 226 (top left, right), 231 (top right, left), 232, 233 (top, center left); Steve Culpepper, ©The Taunton Press, Inc.: pp. ii, 2, 37, 44 (top right), 83, 105, 109, 113 (bottom), 114, 135 (top left), 140, 142, 146 (left), 150, 152, 153 (top), 156 (top left, bottom right), 158 (right), 159, 160 (right), 161 (all except top), 162–163, 166 (left), 167, 169, 170, 172, 177–180, 181 (left), 184 (left), 186–188, 193 (center, bottom), 194 (top, center), 195 (top), 198, 200 (bottom), 201 (top, center), 202, 204, 206, 207 (top), 216 (top), 218 (center right), 233 (right); Bob Falk: pp. 3, 4, 8, 9 (top), 14, 17, 19, 20, 22, 25 (top right), 29, 30 (top, center right and left), 32, 36, 40 (bottom right), 41, 42 (top right, left, bottom right), 43, 44 (top left), 45 (top left, top right, bottom right), 46 (inset), 62 (left), 63 (left, right), 64–65, 70 (top), 74, 75 (top left), 77 (top right), 79 (top left), 80, (top right, bottom), 84 (bottom left and right), 88 (center), 90, 91 (bottom), 92, 95 (top, bottom), 96 (top, bottom), 97, 104, 106 (top), 107 (bottom), 120, 122 (top), 125 (right), 135 (bottom), 138, 144, 145 (all except center), 147, 153 (bottom), 154 (top, center), 155 (top), 156 (top right, bottom left), 157 (top), 158 (top left, bottom), 160 (top, bottom), 161 (top), 164–165, 166 (right), 168, 171, 173, 176, 193 (top), 195 (center, bottom), 196, 207 (center), 210 (top), 216 (center), 238; ©NHTP: p. 15 (bottom); ©Dallas Kline, The Woods Company: p. 16, 23, 34; ©Dick Powell: p. 18, 117, 205, 211 (top right, top left), 223; ©Pete Kreiger: pp. 24 (top left and right, bottom left), 39, 111, 131, 136; ©Brian Vanden Brink: p. 24 (bottom right); ©Scott Lanz: p. 25 (top left); ©Barry Stup: p. 25 (bottom right); ©Lori Cornelia: p. 26; Randy O'Rourke: p. 28; ©Ted Reiff: pp. 40 (top, bottom left), 72 (top), 73 (bottom right and left), 76 (top right), 157 (bottom); ©Beverly Rheaume: p. 46 (top); ©John Janowiak: p. 62 (right), 94, 99, 100; ©Bob Ikens: p. 81 (top); ©Bill Bowman: p. 89; ©George Verlaine: pp. 98, 112 (bottom), 119, 123 (top), 124 (center, bottom), 125 (left), 127 (bottom left), 128 (top, center), 130 (top left and right), 139, 154 (bottom), 155 (bottom), 174, 185, 210 (bottom), 215, 217, 218 (center left, bottom right), 221 (top left), 222, 224 (top right), 226 (bottom), 227–229, 231 (bottom), 233 (bottom left), 234–236; ©Jeff Wagner: p. 128 (bottom right); ©Josh Giltner: p. 129 (bottom left and right); ©Robert Grieshaber: p. 146 (right); ©Lisa McKhann: p. 148; ©Loy Lauden: p. 208; ©Steve Cosper: pp. 211 (bottom right and left), 230

The Taunton Press

The Taunton Press, Inc.,
63 South Main Street, PO Box 5506,
Newtown, CT 06470-5506
e-mail: tp@taunton.com

Editors: Steve Culpepper, Peter Chapman
Jacket design: Scott Santoro, Worksight
Interior design: Scott Santoro, Worksight
Layout: Emily Santoro, Worksight
Flip photos from video courtesy
    Shane Endicott, The Rebuilding Center,
    Portland, Oregon
Illustrator: Chuck Lockhart

Library of Congress
Cataloging-in-Publication Data
Falk, Robert H.
    Unbuilding : salvaging the architectural treasures of unwanted houses / Bob Falk and Brad Guy.
        p. cm.
    Includes bibliographical references and index.
    ISBN-13: 978-1-56158-825-1 (alk. paper)
    ISBN-10: 1-56158-825-3 (alk. paper)
1. Buildings--Salvaging. 2. Building fittings--Recycling. 3. House construction. 4. Construction and demolition debris. 5. Salvage (Waste, etc.) 6. Wrecking. I. Guy, Brad. II. Title.

TH449.F35 2007
690'.26--dc22

                          2006029895
Printed in the United States of America
10 9 8 7 6 5 4 3 2 1

The following manufacturers/names appearing in *Unbuilding* are trademarks: Bungee®, Channellock®, Craigslist[SM], Dacor®, Dumpster®, eBay®, Estwing®, Goodwill Industries[SM], Grizzly®, Habitat for Humanity®, Klein Tools®, Milwaukee®, Nail Kicker®, National Trust for Historic Preservation[SM], Porta Potti®, ReConnx®, Salvation Army[SM], Sawzall®, Sharpie®, Stanley®, St. Vincent de Paul[SM], Styrofoam®, Superbar®, The Extractor™, Transite®, Trex®, Tyvek®, Vise-Grip®, Vulcan®, WD-40®, Wonderbar®

## EPIGRAPH

*"They don't build them like that now," said Harley as he tapped
his wrecking bar against one of the barn's old pegged joints...*

*"There's a right way to do things and a wrong way," he said.
"Then there's the quick way. That's how city folks like to work,
so it costs them most in the long run."*

*The current method of tearing down barns is by bulldozer and
wire cable. But since the old mortised joints are usually stronger
than the whole lengths of beam, when a barn is pulled down in
this way, good wood splits. The roof then falls like a sodden cape
over the whole thing, and demolition becomes harder and takes
longer. So I chose the seemingly slower old-time method;
I started by removing the roof.*
— Eric Sloane, *A Reverence for Wood*, 1965

*Waste...a resource in the wrong place.*
— Chinese proverb

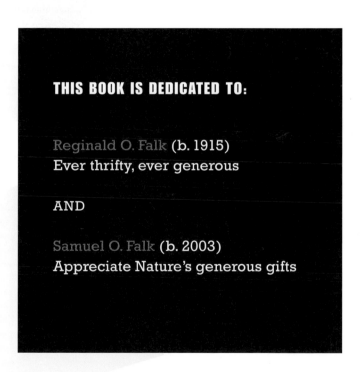

**THIS BOOK IS DEDICATED TO:**

Reginald O. Falk (b. 1915)
Ever thrifty, ever generous

AND

Samuel O. Falk (b. 2003)
Appreciate Nature's generous gifts

# ACKNOWLEDGMENTS

Most successful endeavors require the hands of many. We are grateful to all those who offered inspiration and encouragement throughout the long process of writing this book. Both authors are indebted to Steve Culpepper, executive editor at The Taunton Press. From the beginning, Steve was enthusiastic about this topic and shared our vision for a book on deconstruction and building materials reuse. We also extend our thanks to Taunton Press editor Peter Chapman, for taking us round the last lap toward publication.

Over many years of involvement in the field, we have been inspired by the members of the Building Materials Reuse Association, other practicing deconstructors, and the many others who have helped advance the reuse of building materials, including Peter Yost of 3D Building Solutions; Robin Snyder, Ken Sandler, Pam Swingle, and Timonie Hood of the U.S. Environmental Protection Agency; Jodi Murphy of Murco Recycling; Jen Voichick, Frank Byrne, and Bill Bowman of Habitat for Humanity®; Ted Reiff and Kurt Buss of The ReUse People of America; Jon Giltner of Reconnx, Inc.; Blair Pollock of Orange County Community Recycling; Jim Primdahl (consultant); Tom Napier, Steve Cosper, and Rich Lampo of the U.S. Army Corps of Engineers; Barry Stup of The Woods Company; Pete Krieger and Max Taubert of the Duluth Timber Company; Matt McKinney of the Institution Recycling Network; Bryce Jacobson of Portland Metro; John Stevens of Wood Waste Diversion; Gordon Plume of The GR Plume Company; Pete Hendricks (consultant); Charles Kibert and Kevin Ratkus of the University of Florida; Scott Lantz of the U.S. Forest Service; Stan Cook of the Fort Ord Reuse Authority; Steve Loken of the Center for Resourceful Building Technology; and Jennifer Corson of Resourceful Renovator.

A very special thanks to all the staff at the U.S. Department of Agriculture's Forest Products Laboratory in Madison, Wisconsin, who work every day to foster the wise use of our nation's wood resource.

We are also grateful to everyone who shared photos: Sally Kamprath (ReHouse), George Verlaine (photographer), Chris Wachholz (Cabins, Cottages, and Bungalows), Pete Krieger (Duluth Timber), Adrian Scott Fine (National Trust for Historic Preservation), Jen Voichick and Frank Byrne (Madison Wisconsin Habitat for Humanity ReStore), Ted Reiff (The ReUse People of America), Dick Powell (consultant), Gene Kennedy (Deconstruction Works Community Development Corporation), Loy Lauden (The Whole Log Lumber Company), Barry Stup and Dallas Kline (The Woods Company), John Abrams and Tim Mathiesen (South Mountain Company), Bill Bowman (Habitat for Humanity), John Janowiak (Pennsylvania State University), Scott Lantz (US Forest Service), Bob Ikens (Ikens Hardwood Floors), Robert Grieshaber (University of Wisconsin, Milwaukee), Jon Giltner (ReConnx), Lorie Cornelia (ReHouse), and Lisa McKhann.

Special thanks to Ted Reiff and Gerald Long from The ReUse People of America, Kevin and Brook Horowitz of St. Helena Construction, and Kathy Smith for sharing their Napa, California, deconstruction project.

## FROM BOB

I am grateful to my family for continued inspiration. To my parents, Kay and Reg, who taught me that perseverance, and a little luck, can get you far in life. A big thanks to my two older brothers, Roger (of Falk-Vanderwal Builders, Kalamazoo, Michigan) and Tom (of Stewarts Point Woodworks, The Sea Ranch, California), who helped teach me the finer points of house building and remodeling and an appreciation for the detail found in architecture and antiques. Bros, I wouldn't have had the expertise to write this book without both of you sharing yours. A big bear hug to my Deconstruct'n Cousin, Jim Stowell, of the Whole Log Lumber Company, Zirconia, North Carolina, who shares my love of old wood and who walks the walk every day. Finally, a heartfelt thanks to my loving wife, Grace, and son, Samuel, for their enduring patience with a man who not only takes on too many projects but who clutters their lives with all the stuff he just can't bring himself to throw away.

## FROM BRAD

Thanks for the love and patience of my wife, Lauri Triulzi, who bears with my travels from home and my passion for this work, and my parents, Jeanne and Allan, who showed me that the world is a big place with lots to learn from it.

# CONTENTS

## BOB'S PREFACE

Salvaging building materials started at a young age for me. As the son of a thrifty Depression-era remodeler, my father taught me early on that building materials can live more than one life. If you are not preoccupied with the idea that everything has to be brand new, can use a little ingenuity, and aren't afraid of some hard work, quality building materials can be salvaged and reused, saving lots of money.

Throughout my youth, it seemed we salvaged nearly everything for remodeling projects and house building. Lumber from an old military barracks, insulation from an old ice-storage locker, excess plumbing, used bricks; the list goes on. While at times the thriftiness seemed to go to the extreme (I had to straighten old nails for use in my preteen building projects), I did learn that one man's trash can be another man's treasure.

Salvaging and reusing were so ingrained by adulthood, I didn't appreciate the "resource conser-vation" effort my family had been involved in: We were greening the environment by using fewer mined, logged, and drilled resources; we were reducing landfill impact; and we were supporting a local waste-based business by purchasing from the local salvage yard.

You don't have to go to the extreme of reusing old nails to benefit from "building materials reuse." In our resource-rich nation, there's a wealth of high-quality building materials available for salvage. The windows, doors, cabinets, fixtures, lumber, trim, and hardwood flooring that often end up in the landfill can all be easily salvaged and reused; and these items can be of much higher quality than those found in new construction. Long lengths of hardwood flooring, solid-wood trim and molding, quarter-sawn wood siding, heart-pine stair treads, walnut newel posts, ornate hardware, stained-glass windows, period lighting, elaborate bracketry and trim, and many other items too numerous to list can be found and reused. Many of these items are vintage and aren't available from any other source.

So, how do you go about harvesting this resource, and do you have the skills necessary? If you don't have your own building to remove, there are thousands of buildings torn down every year in the United States, with plenty of opportunities in nearly every community. Check your local newspaper's want ads or place an ad yourself. You can also call the building department and ask about recent demolition permits. Often, a building owner will talk to you if you can remove the building for less than his demolition cost, can meet his schedule, and aren't a liability threat.

Taking apart a building is not unlike putting one together, only in reverse. In many ways it's easier. No measuring, fitting, or cutting to exact lengths or angles. If you have some basic carpentry skills, "unbuilding" and material salvage is very doable. Like construction in general, taking apart a building can be dangerous, so safety is paramount. In this book, we will discuss this unbuilding process, how to remove these materials, and—most important—how to do it safely.

## BRAD'S PREFACE

I can't say that I grew up salvaging. I was raised in the tropical paradise of the southwest coast of Florida, which back then was a pretty sleepy part of the world, hemmed in between the Gulf of Mexico and the Everglades. In the town where I lived, there were no interstate highways, no malls, one high school, one old downtown movie theater, and a long two-lane highway across the seemingly endless Everglades to the big city of Miami. My memories were of the quiet and overwhelming presence of sunlit and lush green nature. I also spent time in my adolescence in North Africa, where, as Anglo-Saxons, we were the intruders in a land of extreme poverty and class disparities. We lived in a walled compound across the street from people living in shelters made from corrugated metal and cardboard. These experiences are what drove me to the field of architecture, with the hope that I might help improve the human environment.

I soon learned that most building destroys rather than rejuvenates nature and does not enliven the spirit. The built environment is a major contributor to the environmental and social problems in the world of today. We face unknown and potentially catastrophic changes resulting from reckless human consumption—global warming, energy shortages, water shortages, even alterations in the human physiology from toxins that have been created and used carelessly. Recovering building materials and reusing them is one small contribution to environmental and social sustainability.

There are many measurable benefits to being resource conservative and following the three Rs of Reduce, Reuse, Recycle. I hope that the notion of unbuilding and the words and photos in this book will encourage you to become more aware of the value of both the unique and the everyday materials in our buildings and to consider the pleasure of reclaiming and giving old things new lives.

# IS UNBUILDING FOR YOU?

## Introduction

*Dig where the gold is...unless you just need some exercise.*
—John Capozzi

This book is written for anyone interested in the salvage and reuse of building materials. We believe there are buried treasures in many of the buildings that are torn down every day; and whether you are a do-it-yourselfer, a professional builder, an architect, or a homeowner interested in using reclaimed building materials in your own construction project, we've written this book to help you find those treasures.

While we give you enough information to safely deconstruct a wood-framed building, we also provide useful facts on the characteristics, qualities, and reuse potential of materials salvaged from such an endeavor.

*Americorps National Civilian Community Corps members deconstruct a military warehouse building at Fort Campbell Army Base. This building yielded over $32,000 in reusable materials.*

## How to Use This Book

Taking apart a building is not unlike building one, only in reverse. However, no measuring or fitting of parts is necessary, so disassembly can be surprisingly fast when compared to putting a building together. Similar to residential construction, deconstruction requires hand labor. Most remodeling involves some selective hand demolition, so most remodelers are familiar with the basic process and only small changes need to be made to recover the "waste" from renovation projects.

As with most do-it-yourself building projects, the savings reaped are in proportion to the sweat expended. Although salvaging these materials can sometimes be dirty and exhaustive, the items recovered from old buildings can be of much higher quality than those available for new construction.

This book has been written so that the information can be easily accessed. Chapter 1 offers a general discussion of building disassembly and materials salvage. Chapters 2 through 4 are dedicated to the planning and organizing of an unbuilding project, focusing on the important questions to be asked before getting started as well as considerations of logistics, labor, and site evaluation. Chapter 5 focuses on safety and health issues. Finally, chapters 6 and 7 describe the actual process of building materials salvage, including both "soft-stripping" (cherry-picking the most prized material and fixtures) and full deconstruction.

Throughout the book we also feature profiles of individuals who might be considered "unbuilding pioneers." This diverse group of entrepreneurs has proven that opportunities abound across our nation in harvesting the wealth of materials found in existing buildings.

### The Benefits of Unbuilding

- High-quality, reusable materials can be found in nearly every building.
- Basic building materials can be provided at lower costs than by buying new.
- Unbuilding is much gentler on the immediate environment than demolition, avoiding the potential for noise; dust; and damage to soils, plants, and nearby buildings.
- Fewer materials go to the landfill.
- Unbuilding helps extend the earth's natural resources because every pound of material reused is more than a pound of extracted resources that doesn't have to be manufactured new.
- Unbuilding helps preserve historic architectural features that might otherwise be lost.
- Because there are so many buildings slated for removal in our urban areas, unbuilding has the potential to create jobs where they are needed most: in our inner cities.

*Deconstruction* is an environmentally preferable alternative to smashing a building and landfilling it.

As with any building construction project, safety is paramount and we discuss the methods and gear to keep you safe and healthy on the job site. We provide many important labor- and time-saving tips that will make the salvage and reuse of building materials as easy as possible.

## What Is Unbuilding?

Unbuilding, or building deconstruction, is gaining popularity as an environmentally preferable way to save and reuse building materials. Unbuilding can be thought of (loosely) as reverse construction and is generally perceived as manual disassembly of a building, although various combinations of manual and mechanical methods are used to improve cost and time performance. Very high material recovery and reuse rates are possible.

As an alternative to smashing a building down and sending the debris to the landfill, deconstruction is suited to recovering materials for reuse. Because there are many materials that are difficult to recover or that are not appropriate for reuse (broken wood, worn asphalt shingles, hazardous materials such as asbestos), deconstruction can also be beneficial because it entails more careful separation, handling, and disposal of these materials. In addition, materials that are not reusable but that are recyclable are more easily separated out during the unbuilding process.

## Who Can Do It?

You don't need to be a professional builder or have a mastery of carpentry to use this book. Nor do you need a stockpile of woodworking tools. If you recognize that many used materials can work as well as new ones, have a measure of patience, aren't afraid of physical work and getting a little dirty, you can salvage and reuse hundreds of quality building materials. For your effort, you can save money and acquire materials that are often better than new.

## Why Unbuild?

Depending on your generation, you may have heard proverbs such as "Waste not, want not" and "It is thrifty to prepare today for the wants of tomorrow." We certainly did; and although these sayings summarize the motivation for many of us, there are many other incentives for unbuilding and reuse. Thrift is certainly a prime motivator, but a desire to be environmentally sensitive, a fondness for antiques and other items from the past, the yearning to have more control over the quality of materials

*No special training* is needed to salvage high-quality materials from buildings.

used in construction, and the recognition that many of the materials available for salvage are of higher quality than those produced today are all valid reasons.

As with many environmentally conscious activities, deconstruction and building materials reuse offer a direct and measurable way to reduce one's negative effect on the planet. Building construction, use, and maintenance make up a resource-intense business. In the United States, the construction, use, maintenance, and disposal of houses are responsible for nearly a half of the country's energy use. According to the U.S. Geological Survey, about 60 percent of *all* materials (except food and fuel) used in the economy each year is consumed by the construction industry.

Certainly there are many opportunities to reclaim and reuse building materials. In 1996, the U.S.

Environmental Protection Agency (EPA) estimated that the equivalent of 250,000 single-family homes are disposed of each year, which represents an estimated 1 billion-plus board feet (bd. ft.) of available salvageable structural lumber, or about 3 percent of our annual softwood timber harvest. Reusing this lumber could save 4,250,000 trees on 150,000 acres of timberland every year. The amount of recoverable materials is even greater if you add nonstructural building products, such as the millions of windows, doors, and fixtures and the thousands of miles of trim work, siding, and flooring available.

Not only does unbuilding (and the reuse of building materials) save resources but it can also yield higher-quality materials than are available today. Much of the salvaged lumber available through deconstruction

*Demolition and landfill of reusable building materials wastes our nation's resources.*

is from the decades of old-growth harvest (higher density, slower grown, fewer defects), which represents a resource largely unavailable today. In old factories, silos, and water tanks you can find high-quality heart pine, redwood, and fir timbers; in old barns, pine, chestnut, and oak; and in older school bleachers and benches, quality maple and fir. Factories, farms, and industrial buildings aren't the only sources of such materials; high-quality wood can also be found in the millions of older homes across our nation. In nearly every community, wood flooring, windows, doors, cabinets, and lumber can be salvaged. And if you keep your eyes open, high-quality architectural materials, including hardware, period lighting, elaborate bracketry, and trim, are also readily available.

*Many abandoned buildings* have the potential to be deconstructed and the building materials salvaged, such as these row houses in Philadelphia.

*Natural disasters can* provide a source of salvageable building materials. These high-quality cypress timbers were salvaged from buildings slated for demolition after being damaged by hurricane Katrina.

## Reuse vs. Recycling

Many of us have heard the three Rs of waste reduction: Reduce, Reuse, Recycle! This hierarchy suggests that to be as environmentally benign as possible we should first reduce our level of material use, then reuse as many materials as possible, and finally recycle what can't be reused.

Unbuilding focuses on the reuse and recycling portion of the three Rs mantra, and it's important to make a distinction among these activities. Reuse is the heart of the deconstruction effort, by which the primary focus is to maintain the material or component in its original form. (This might include some cleaning, removal, or replacement of some part.) The intent is to move these salvaged building materials *directly* back into new construction or remodeling: a used door in place of a new door, a used window instead of a new window, a salvaged 2×6 rather than a new 2×6, and so on. Or the salvaged materials can be used in a new way: a used door becomes a wall panel, a window serves as a cabinet front, and the 2×6 floor joist is now a 2×6 wall stud. Recycling, on the other hand, is a more *indirect* use of materials and typically involves changing the form of the material for use as an entirely new material. For example, we wouldn't reuse a concrete pillar from an old building in a new building. It's not practical. However, it is very practical to break up that pillar and recycle the concrete and steel rebar into other uses, such as in roadbeds and new cars, respectively.

Unbuilding and its associated reuse are very well suited to *wood-framed* construction, where most materials can be reused. Demolition and its associated emphasis on recycling are well suited to *concrete and steel* construction, whose materials are difficult or impossible to directly reuse, and breaking down these materials is an inherent part of the recycling process.

## Unbuilding Is Not an Alternative to Historic Preservation

We hope you consider deconstruction only when building preservation or adaptation is not an option. Most buildings are not historically significant; however, it's best if you determine that before deconstruction. Historic preservation typically is invoked only to protect a significant historic, archaeological, or cultural resource. We suggest that when preservation of the whole structure is not possible, unbuilding can *at least* serve as "preservation in pieces" to recover significant construction materials and design features. In some cases, a building shell or only the street front will be retained to preserve historic building character in a community, while the rest is removed to allow for a modern interior. Given the selective nature of these projects and the care needed to avoid damaging the parts that might remain, unbuilding is a viable solution. In these situations, the original materials can often be directly incorporated in the preserved building.

*Old buildings* are a rich source for construction materials and architectural items, from doorknobs and sinks to stained glass.

## Unbuilding as a Green Endeavor

Unbuilding is the ultimate green endeavor and the first step in preserving resources and avoiding waste; it is directly in line with the tenets of "green building." A simple raised wood-floor, wood-framed older house can weigh 50 lb. per sq. ft. A 1,500-sq.-ft. light wood-frame building can therefore weigh more than 37 tons or the volume of about three 40-cu.-yd. container loads. In addition to these raw materials of the building, there are also the original materials consumed to make the finished building materials and the energy and pollution resulting from extraction, processing, and transportation at different stages of manufacture. The waste itself becomes a burden in the landfill, consuming land and potentially leaching into soils and groundwater. By deconstructing, you are also doing your neighbors the very simple favor of extending the life of your local landfill.

The United States is one of the most consumptive nations on the planet. According to John Ryan and Alan Durning's book, *Stuff: The Secret Life of Everyday Things*, the average citizen consumes about 120 lb. of natural resources each day. With only about 5 percent of the world's population, the United States consumes about 24 percent of the world's energy. Unbuilding can help decrease this percentage through the direct reuse of building products.

With a greater realization of the positive effects of greener building practices, the use of recovered materials is typically rewarded in the many green building certification programs that have appeared in recent years. Organizations such as the U.S. Green Building Council, the National Association of Home Builders, and the National Association of the Remodeling Industry have all developed green building programs for use by both the commercial and the residential building market. A green building certification provides not only a perception of quality but also real energy savings and increased value on a new or renovation project over the long term. More and more municipalities are considering policies and ordinances requiring building deconstruction to be considered along with traditional demolition and disposal.

*Your local reuse store can offer a cornucopia of reusable building materials.*

# UNBUILDING OPPORTUNITIES

## Chapter 1

> *Opportunity is missed by most people because it is dressed in overalls and looks like work.*
>
> —Thomas A. Edison

The opportunity to unbuild can present itself in a variety of ways and forms. Although many of these opportunities are obvious, others are not; and we offer here some suggestions on finding a deconstruction project that meets your needs. In addition, we talk about building variety, material types, and construction practices and how they can affect your decision to unbuild.

## Redevelopment

Certain areas of the country are in a frenzy of old house demolition to make way for new buildings. Studies show that land reuse is the most common reason for demolitions. Usually in these cases, the land has become more valuable than the building that sits on it and a decision has been made to remove the building to put up another, either residential or commercial, that is larger, more functional, of greater value, or of more appropriate use. Redevelopment is providing many opportunities for deconstruction. Older homes often have architectural treasures covered by years of renovations. And newer homes that are torn down, though lacking

*Many architectural items* were salvaged from this turn-of-the-century Madison, Wisconsin, house before it was demolished.

Teardowns are increasingly common in older neighborhoods.

*A common result* of a teardown is a rebuild that is out of scale and out of place with the existing older-style homes in the community.

in period architectural elements, are often in better shape and are unlikely to have hazardous materials, such as lead-based paint and asbestos.

Unfortunately, there is an increasing trend in older, often historic neighborhoods to demolish smaller, older homes to construct a larger one on the same lot, a trend referred to as "teardown." This trend has gotten so extreme that homes less than 10 years old have been torn down to make way for bigger ones.

## The Teardown Epidemic

In two neighborhoods just outside downtown Dallas, more than 1,000 early-20th-century homes have been bulldozed and sent to the dump, making way for the construction of 10,000-sq.-ft. luxury homes. In Denver, some 200 homes, most of them brick bungalows from the 1920s and 1930s, were demolished and replaced with stucco-clad houses three times their size. In Rancho Mirage, California, a 5,000-sq.-ft. museum-quality home designed in 1962 by famed architect Richard Neutra was torn down without warning by its new owners, who plan to build a much larger new house.

These are just three examples among many of a disturbing pattern of demolitions that's approaching epidemic proportions in historic neighborhoods across America. The National Trust for Historic Preservation[SM] has documented more than 100 communities in 20 states that are experiencing significant numbers of teardowns.

Although unbuilding is not a solution to the teardown trend, it at least offers a means to preserve for reuse the materials and architectural elements of the lost buildings —materials that would otherwise end up in our already overcrowded landfills.

## Rural Property

Barns and other outbuildings on rural properties present some of the best opportunities for unbuilding. Many old American barns have become dilapidated over the last several decades and increasingly are candidates for demolition. Businesses have started up that specialize in barn dismantling and reuse, and certainly these structures can provide a wealth of wood materials. There are also specialty companies that focus on dismantling barns to reassemble and upgrade them as fine houses or for commercial use. This is the best kind of unbuilding—deconstruction for the purpose of preserving and building anew.

## Military Bases

A large number of existing military facilities were built during World War II, when steel was in great demand for the war effort and construction was wholly or partially from wood. As a result of base closings and downsizings, thousands of buildings are currently slated for disposal. These range from industrial buildings several hundred thousand square feet in size to the ubiquitous one- and two-story army barracks.

*Though many barns* are disappearing from the American landscape, they can be a source of high-quality timber.

Although not every surplus building in the military is available, or suitable, for deconstruction, more and more are being deconstructed. The military has clearly recognized that more environmentally preferable options exist than "smash and bury."

As an example, the army's Fort McCoy in Wisconsin has adopted a standard practice of selling surplus World War II–era buildings through competitive bidding. Since the early 1990s, they have overseen individuals, families, and small building contractors who have deconstructed more than 140 buildings. The successful bidder signs a contract and makes payment to the U.S. Treasury. Fort McCoy removes all asbestos and hazardous materials before the buyer begins work. Lumber is typically the most valuable and most sought-after material.

Unbuilding has been good for:

→ the successful bidders, who have used the materials salvaged to build homes, garages, and churches;

→ the environment, which receives less waste;

→ taxpayers, who are saved the cost of demolition and landfill tipping fees and are rewarded with the income received from bidders.

The staff at Fort McCoy has concluded that average building removal costs have been reduced from $40,000 per building using a standard demolition contract to $2,000 to $5,000 per building using deconstruction.

*More than 1,600 buildings are slated for disposal at Fort Ord, a closed U.S. Army base in California.*

*Even the most mundane*-looking military buildings can contain beautiful lumber. These munitions storage buildings at the U.S. Army's Badger Army Ammunition Plant in Wisconsin are slated for demolition and are good candidates for deconstruction.

## Urban Renewal

In many inner-city areas, abandoned and deteriorating buildings are being torn down by the thousands. Often these older buildings contain not only reusable lumber and timber but also architecturally valuable items. These situations provide an opportunity to conduct salvage, though working with a municipal government typically requires licensing, bonding, and insurance. If you live in a large city, you might contact the urban planning department and see what opportunities exist. We know of situations in which the local government will make agreements to allow predemolition salvage, particularly by nonprofit groups. Removal of these materials can also be helpful in creating economic opportunities, with the reuse of the materials going toward repairs and renovation elsewhere in the same neighborhood.

*There are thousands of condemned housing units in American inner cities, such as this one in Philadelphia. These buildings depress property values in neighborhoods where jobs are sorely needed. Deconstructing these buildings can keep the historic materials in the communities and assist in local economic development at the same time.*

## Unbuilding Philadelphia

The City of Philadelphia Neighborhood Transformation Initiative (NTI) is a comprehensive 5-year plan that addresses urban blight through a multi-faceted program to preserve and build healthy communities throughout the city. One important goal of the program is the demolition of several thousand housing units that are structurally unsound or otherwise unfit for habitation. NTI recognizes the value that still exists in the condemned buildings and is exploring ways to cost-effectively recover salvageable materials. The key issues, as with many places, are the labor costs to remove the materials and the creation of local markets for these goods.

Like many cities, such as Cleveland, Detroit, and smaller Rust Belt communities, Philadelphia shares the unfortunate circumstances of economic and population decline. If the building infrastructure cannot be maintained, it might be considered a kind of merciful treatment to unbuild before too much decay sets in, so that the building materials are retained and pieces of architectural heritage can be put to reuse.

During the remodeling process, every remodeler has performed the rudiments of unbuilding. Although many remodeling jobs are small and the volume of salvageable materials is rather low, opportunities still abound. The current popularity of kitchen and bath remodels can yield many reusable products and materials, some only a few years old. If you are looking for materials for your own remodel, it can sometimes be fruitful to talk to remodeling contractors who work in higher-end neighborhoods.

Recently, a friendly conversation with a contractor led to an opportunity to salvage materials from a kitchen-family room-bath remodel in an expensive home. The owner wanted to replace all existing cabinets, appliances, and so on. A long week-end of work resulted in a kitchen full of solid-cherry cabinets, a family room of built-in cherry bookshelves, a marble-top island, Dacor® appliances (some less than 2 years old), plumbing fixtures, and a $3,000 jetted bathtub. The owner was happy because this salvage operation saved him the $800 he would have been charged for removal; the contractor was happy that he was saved two days in a tight schedule; and the salvager was pleased as punch for trading a few days' labor for more than enough materials to do a complete kitchen remodel.

If you find yourself planning a renovation of your own house, then you might consider doing the deconstruction yourself or with your contractor. If the materials fit back into your renovation plans, you might save yourself a few dollars by reusing them in the new construction work. If you don't need the materials and there is a local nonprofit building materials reuse center in your area, you may be able to donate the salvage materials and get a tax-deductible donation receipt for tax purposes. In either case, you will keep the materials out of the landfill.

*Older homes* aren't the only place to look for unbuilding opportunities. High-quality items such as these cherry cabinets can be salvaged from newer remodels too.

## Building Auctions

If you don't have a building to deconstruct, another way to obtain used building materials and do the work yourself is through on-site building auctions. Two companies that specialize in these auctions are Murco Recycling, in LaGrange, Illinois, and Whole House Building Supply and Salvage, in East Palo Alto,

California. Others are sure to follow. Murco Recycling works solely as an auctioneer/broker, whereas Whole House has retail facilities for selling items that go unsold at auction. Both advertise the sales, manage the on-site process, and maintain email lists of potential customers who are kept up to date on pending auctions.

*An on-site building materials auction can draw a big crowd, with everyone eager to find a bargain.*

*Wisconsin Habitat for Humanity crews deconstruct a surplus army building.*

As with any other type of auction, buyers bid on items and typically must remove everything the same day. Real bargains can be found at these auctions because they usually have lower prices than a retail used building material store since the costs of removal and transportation are borne by the bidder. Bidders must bring their own tools, help, and a vehicle capable of hauling off their load. Jodi Murphy, owner of Murco Recycling, says, "It's amazing what you find—brand-new kitchens, brand-new appliances, brand-new windows, granite countertops—all in houses getting torn down."

## Habitat for Humanity ReStores

One of the biggest trends in the unbuilding world is the growth of Habitat for Humanity ReStores, with at least one store in nearly every state. Although many ReStores primarily obtain like-new materials from commercial donors, many also conduct some degree of deconstruction and salvage. Just as Habitat uses volunteers to build affordable houses, these ReStores use volunteers to conduct salvage and deconstruction projects. Volunteering to help out at a local ReStore can be a great way to get your feet wet in unbuilding and possibly give you a first look at some tools and materials you may want for yourself. You can contribute your labor or, if buying salvaged materials from the ReStore, help fund the construction of new affordable housing.

## What to Unbuild

The types of buildings available for unbuilding vary tremendously in design and construction. How a building was constructed can tell you a lot about the materials you can expect to find, how easily it will come apart, the best approach and sequence to follow, and the safety and health issues you face. Figuring out

the details of a building's construction and materials can take a little extra effort if it has been subjected to multiple renovations. A building can be easy to deconstruct or difficult, depending on when it was built, its original function, degrees of alterations, and the materials used.

Don't be surprised to see strange and inconsistent construction and materials, even from wall to wall and room to room. A simple project can be complicated by added layers of low-value materials, which will end up as waste or recyclables and need to be removed before the harvest of reusable materials can begin. This overburden may include unsalvageable wall, floor, and ceiling materials; utilities, including ductwork, piping, and electrical wiring or conduits; or simply the furnishings and trash from previous occupants or squatters. A critical element of this overburden is hazardous materials (see chapter 5 for how to deal with these materials).

Most structures suitable for deconstruction are wood framed. Because concrete and steel are typically recycled rather than reused, our focus in this book is on wood-frame construction. The three most common building types of interest are timber framed (light and heavy industrial), barns, and single-family wood framed.

## Heavy timber framing/light-frame commercial

Timber framing is one of the oldest forms of wood construction. Also called post-and-beam framing, this traditional system relies on a skeleton of heavy posts (vertical members) and beams (horizontal members) to frame the structure. Areas between the framed sections were infilled, often with brick, to form the exterior and interior walls. Many industrial buildings built in the 19th and early 20th centuries were timber framed.

*Now you see it, now you don't. This old barn in New Enterprise, Pennsylvania, yielded a rich source of framing lumber.*

*Timber-frame barns and old factories are great sources of lumber, but the deconstruction work can be challenging.*

However, whereas early post-and-beam frames were held together by mortise-and-tenon joints, the greater availability of metal hardware in the late 19th century resulted in the use of bolted connectors or other metal hardware to fasten the frame together. A movement back to hand-constructed joints is evident in some contemporary timber frames, many of which use reclaimed timbers from unbuilding.

Deconstructing a timber-frame industrial building requires a different level of involvement than you'd need in a single-family home. On the one hand, a large industrial building requires heavier equipment to remove and handle timbers, and more storage and work space; on the other hand, the yields of quality heavy timber can be high. This work also requires a significant amount of experience, equipment, insurance, bonding, and time. It is not for the faint of heart!

You may find lighter-frame commercial and industrial buildings available for deconstruction that can yield a lot of materials. Often, the construction is similar to that used in a single-family home. Most are built with nothing bigger than a 2×12, with individual pieces of lumber nailed together to build up beams and columns. In many older structures, you'll find the lumber bolted or spiked together to create large-span trusses.

*The heavy timbers found in old timber-frame buildings can be reused in contemporary construction.*

This **600,000-sq.-ft.** *military industrial building dating from World War II yielded over 1 million bd. ft. of heavy timber.*

**The wing** *of this commercial storage building was stripped of its metal exterior, overhead doors, and wood sheathing, and the materials were later used to build two new residential garages.*

## Barns

If timber is what you're after, barns offer a much more manageable deconstruction project than a large industrial building from logistical, permitting, and liability standpoints. Older barns were often constructed of softwood timbers, though hardwoods such as oak and chestnut can be found in some parts of the country. Depending on the age of the barn, these timbers might be hand hewn; this material is popular for reuse as exposed rustic timbers in new construction. Both the timbers and the barn siding (if it's thick enough) can be remilled into other building products, such as flooring, trim, and stock for furniture. Barns are often large structures with big beams and posts and may require heavy equipment to safely deconstruct; however, the fact that they are on rural property means that there's usually plenty of room to work around them.

*The structural timber framing, the siding, and the roofing (if tin) can all be salvaged from barns.*

Kathy Burdick (left) and Sally Kamprath (right) own ReHouse Inc., in Rochester, New York, and salvage usable building materials for resale. They often use on-site auctions to market their finds.

"Some of our customers can't believe what is put on the curb for garbage pickup."

# Going, Going, Gone!

## Sally Kamprath
## Kathy Burdick

For Sally Kamprath and Kathy Burdick, abandoned houses awaiting demolition are hidden treasures full of value. Sally started ReHouse Inc., in Rochester, New York, in 2002 after seeing many properties bulldozed without any effort to salvage the usable materials. ReHouse began as an auction business, conducting onsite sales of used building materials from homes slated for the wrecking ball.

To get started, Sally searched local planning board minutes and asked around to find houses scheduled for demolition. "I contacted the property owners to work out a deal to auction off as much of the salvageable materials inside the house as possible. In exchange, we gave the property owner a percentage of the auction proceeds. And they also saved dumping fees at the landfill."

Word got out, and the auctions were well attended. Sally's auction-based business model kept costs down because there was no need to carry an inventory and no need for rental space, but Sally couldn't help thinking she was missing a bigger opportunity. It became clear that the way to recycle more was to open a reused building materials storefront that would be accessible to more of the community.

Sally attended the fall 2004 Deconstruction Conference held by the Building Materials ReUse Association (BMRA) in Oakland, California, and came away with the information and confidence needed to go the next step and open the retail store. David Bennink of ReUse Consulting of Seattle, Washington, was hired to help with the setup of the new operation.

Kathy Burdick came on board in 2004 after hearing Sally on a local radio station explain the salvage auction business and her plans for expansion. They found a 12,000-sq.-ft. warehouse and opened the ReHouse store. The building is packed with an eclectic mix of items—some modern, some older, and some antique. ReHouse offers everything from windows, doors, toilets, sinks, wardrobes, lumber, furnaces, appliances, and entire kitchen units to smaller items like ornate doorknobs and rosettes, light fixtures, and cabinet hardware.

Given that inventory is not something that can be ordered up, they are continually in search of materials. ReHouse employees have the skills to salvage whole houses, and they also receive inventory from landlords who are upgrading their properties and homeowners who are renovating their houses. "Some of our customers can't believe what is put on the curb for garbage pickup and will actually fetch the items themselves and deliver them to us to sell in the store. Without an outlet, these items would clutter a landfill and would be lost forever. The landfill's loss is the city's gain."

ReHouse is Rochester's only reuse store. "Though the creation of a new business is hard work and the hours are long, the rewards are many. We can go home at night knowing we are making a difference in our small corner of the world."

**Balloon vs. Platform Framing**

From a deconstruction standpoint, a balloon-frame house requires a slightly different approach than does a platform-frame house (as we will see in chapter 7). In a platform-frame house, you can typically pull the second-floor walls onto the second-floor deck for further disassembly. In a balloon-frame house, the full-height studs will have to be cut just above the second floor, unless they are to be kept full length, in which case more effort is required to completely separate the wall and floor framing in place.

## Single-family homes

The construction of the single-family home has changed significantly since World War II. Because you're more likely to deconstruct an older home than a newer one, it's important to understand how the materials and construction techniques have changed over time so you know what you might run into in a particular building.

Our earliest homes were built from logs. As sawmills became more prevalent, sawn timbers were used in timber-frame construction. The light-frame construction techniques that we use today have evolved over time and replaced timber framing and log construction when uniformly sawn lumber and cheap metal fasteners became widely available.

### FRAMING

You'll find two basic types of framing in single-family homes: balloon framing and platform framing. Balloon framing is an older style of framing in which the studs in the walls run uninterrupted from the sole plate or foundation to the top plate of the second floor; floor joists are nailed directly to the side of the studs. The disadvantage of this system is the tendency for the open stud cavities to allow a fire to travel rapidly to the upper floors and attic, which is why so-called *fire stops*, or short horizontal members between the studs, were installed. Balloon framing could be erected faster than a post-and-beam frame and could be built by less-skilled labor. Since World War II, platform-frame construction has evolved as an even more efficient method of house building.

Most modern houses are platform framed, in which a single floor of walls is erected and decked with joists and sheathing that create a platform to construct the second and succeeding floors. These shorter platform-frame walls are easier to build and don't require the long pieces of lumber used in balloon framing. The walls can be fabricated and erected on each floor platform, which increases safety and reduces labor cost. No added fire stops are necessary because each floor platform encloses the stud spaces above and below.

*In a platform-framed house,* each floor of walls is constructed one at a time.

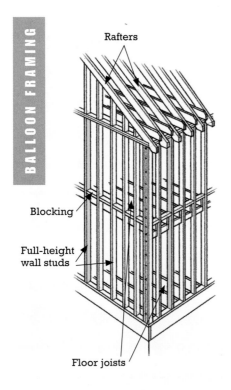

Rafters

Blocking

Full-height wall studs

Floor joists

*Older, raftered roof systems are typically constructed with larger-dimension, longer-length lumber, which can offer rich rewards for the unbuilder.*

*Today, most house roofs are constructed with metal plate–connected trusses, which may or may not be reusable.*

## RAFTERS AND TRUSSES

Before World War II, most houses were constructed using rafters for the roof structure. Depending on span, these rafters ranged from 2×4s to 2×12s. A center ridge board (often 1×) was used to tie the rafters together at the ridge. Today, most house roofs are built using prefabricated trusses or prefabricated I-joists. Using shorter, and often smaller, pieces of wood, both trusses and I-joists are efficient structural systems.

In terms of salvage, rafter roofs will yield more reusable lumber than trusses; however, trusses can be salvaged whole and, depending on the number available of the same span, may be practical for reuse. Check with your local building department to make sure truss reuse is allowed. As in new roof construction, dealing with a system of trusses is a tricky business because trusses become very unstable when the bracing between them is removed. A whole system of trusses can easily topple if the end or intermediate bracing is removed.

## SHEATHING

Plywood, developed early in the 20th century, underwent a boom in production after World War II and quickly replaced the solid 1× boards that had been commonly used for wall and roof sheathing. Today, oriented strand board (OSB) has largely replaced plywood as the sheathing of choice.

Salvage of 1× sheathing, especially roof sheathing, is often questionable because of the holes from the many nails used to attach the roofing. If the building was reroofed several times, the sheathing can look more like Swiss cheese than lumber. Over time, the roofing felt can also melt and leave a sticky coating on the sheathing, and the heat of the sun can dry out the wood and make it very brittle. Each roof must be inspected to determine the amount of damage, and the practicality of salvage will depend on local demand for 1× sheathing or your specific reuse plans. The salvage of plywood and OSB sheathing from newer homes is also possible.

*Oriented strand board* (OSB) *is the most commonly used sheathing in new construction, but may not be practical to salvage.*

## WOOD SIDING AND TRIM

Although the 1× roof, floor, and wall sheathing used in many older homes was typically a lower grade of lumber, wood siding and trim was wood of the highest grade. The size of defects and knots was very limited compared to construction lumber because the siding had to prevent water intrusion, look good, and hold paint. Do not ignore the wood siding on an older home. Wide, long, and surprisingly clear lengths of cedar, redwood, cypress, or pine can be salvaged. Many people simply flip over the siding and reapply with the un-painted back face to the exterior (after trimming, of course). Siding has to be carefully removed because it can be very dry and splits easily.

## LATH AND PLASTER

Though almost all homes are finished with drywall these days, the interior walls of most homes built before World War II were constructed with wood lath and plaster. Spaced wood strips (called lath) were nailed horizontally to the framing, and wet plaster was applied onto the lath in layers. As plaster was applied over the lath, it was pushed into the spaces between the lath, forming a "key" to hold the plaster in place. A thick brown coat was followed by an equally thick coarse scratch coat, which was covered finally by a thin white finish coat. This all added up to quite a bit of weight and a very rigid surface. Some houses also used metal lath. Lath and plaster are probably some of the messiest and most labor-intensive materials to remove, though longer pieces of lath can come in handy to use as stickers for stacking recovered lumber.

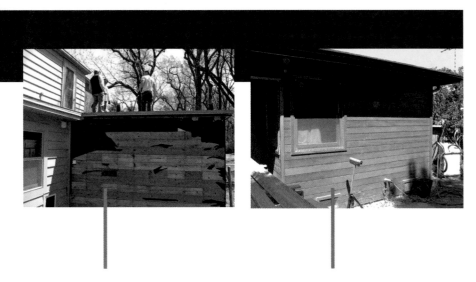

*This garage* was taken down as part of a partial house remodel. The siding, framing, and solid-board sheathing shown here all have salvage potential.

*The painted redwood* beveled siding on this house was carefully removed, flipped over, and reinstalled to highlight the natural wood shown here.

*This wall* in an older home shows the wood lath after the plaster has been re-moved. The keys holding the plaster in place on the wall on the other side are visible between the lath at right.

## Brick construction

Brick was prevalent in the past as an exterior finish material in certain parts of the country. In many urban areas, brick construction is also used for structural walls, especially in common-wall buildings, such as row houses. The bricks were typically laid up at 90-degree angles to each other and overlapped, so that they formed an interlocked wall. It is common to see the interior core of the brick wall made up of "salmon" brick, a pinkish colored brick that was softer, more porous, and lighter colored than the harder and more durable brick of the exterior faces. In older brick structures, wood strips were typically inset into the inside layer of brick to create a nailing surface for the interior finishes. This interior brick is usually not worth salvaging.

Exterior brick can be an excellent material for reuse in new construction. Brick construction before World War II often used lime-based mortar, which is softer and easier to remove than postwar Portland cement–based mortars that adhere tenaciously to the

bricks and can be difficult to remove. Depending on the pedigree, some brick can be very valuable for reuse; for example, Chicago brick can be found reused in places as far away as Miami and California.

## Other materials

More and more houses, especially in the South and Southwest, are built from masonry and light-gauge metal framing. Vinyl siding, pressure-treated wood, engineered I-joists, OSB, and composites of wood fiber-plastic such as Trex® are all recent material innovations found predominately in houses built in the last couple of decades. More recent alternatives like aerated autoclaved concrete block, insulating concrete form (ICF) systems, and structural insulated panels (SIPs) are becoming more commonplace; it remains to be seen how adaptable these systems are to salvage and reuse.

*Bricks are well worth salvaging; however, your time is better spent on solid brick. Three-hole brick is typically not worth salvaging.*

*Jen Voichick started the Madison,
Wisconsin, Habitat ReStore in 2001.
It was the first in the state of Wisconsin
and has served as a model for several
new Midwest ReStores that have
opened with Jen's help.*

"My vision is that people continue to
reuse products so that 'throwing away' is
not such a common occurrence."

# A Big Crazy Idea:
# Start a Habitat for Humanity ReStore

**Jen Voichick**

Though Habitat for Humanity (HfH) has been building homes for low-income families since 1976, it wasn't until 1992 that the first HfH ReStore opened its doors. HfH ReStores typically accept donations of new, overstocked, and used building materials from businesses, suppliers, and homeowners and sell these goods to the public to benefit HfH's homebuilding efforts. The Madison, Wisconsin, ReStore is a good example of a successful store and of how a commitment to helping those in need, a supportive city, and some luck helped create a community-based nonprofit business.

Jen Voichick had the right stuff to start the Madison ReStore, though it took a less than conventional career path to get her there. After college, a stint as a player on the U.S. women's ice hockey team, and a couple of years' apprenticing as a carpenter, Jen started working as a supervisor for a community program working with at-risk youth building affordable houses.

Unfortunately, an injury sidelined her carpentry career, and she was left wondering what to do next. "I kept thinking about the satisfaction I had in building houses for the needy, and at the same time I felt a desire to give something back to the earth to honor the joy and wonder I have encountered in the outdoors. Developing a community-oriented business that supported affordable housing while encouraging recycling and reuse made perfect sense to me."

When Jen first had the idea of opening a nonprofit salvaged building materials store in Madison, she wasn't aware that HfH ReStores existed. In fact, she made a presentation at a local HfH affiliate board meeting only to be politely informed that there were many other stores like this all over the country and that they were called Habitat ReStores. "Boy, talk about having my bubble burst!"

"Though I wasn't the first to think of such a fantastic matchup, I was fortunate that others had already laid the groundwork for me, and I am lucky to be working for something I truly believe in. There are few people who have a big crazy idea and actually go and get the town to pull together to make it happen."

Jen's planned opening date came at an uncertain time, just days after the September 11 terrorist attacks. "I consciously decided to go ahead with the opening because I believed the opening of the store was a small symbol of hope for our world. My vision is that people continue to reuse products so that 'throwing away' is not such a common occurrence."

The Habitat ReStore in Madison carries building materials such as windows, doors, cabinets, lighting, tile, carpet, and hardware; donations come from homeowners, building contractors, building material suppliers, and the store's building deconstruction team. "We are very fortunate in Madison to have such a dedicated, talented, and fun volunteer workforce," says Jen. "The people who work here truly believe in the mission of HfH, and the dedication and community spirit apparent at our store is something that has promoted goodwill—and our world can certainly use more of that."

# DECIDING ON UNBUILDING AND SALVAGE

## Chapter 2

*Ever notice that "what the hell" is always the right decision?*
—Marilyn Monroe

*When evaluating a structure for unbuilding, the first concern should always be safety.*

When evaluating a building for deconstruction and salvage, there are four basic things to bear in mind. Your first concern should always be health and safety. Though diving into an unbuilding project may require a certain level of adventurousness (as Ms. Monroe suggests), no amount of salvaged material is worth the risk of serious injury or compromised health because of unsafe work practices or exposure to hazardous materials. Second, you should think about how extensive a salvage project you want to take on. Do you just want the easy-to-get architectural materials or do you want every last 2×4 in the building? Third, what are you going to do with the materials once they are removed from the building?

Will they be used in another project or will you try to sell them? And fourth, where and how will you store the materials and how will you handle debris?

As with any construction project, a little forethought and planning can save a lot of headaches later on. In this chapter, we provide an overview of going through a building before you get your hands dirty to evaluate if a project is worth pursuing—from safety, health, and logistic standpoints.

Hazardous materials abatement is a consideration for any unbuilding project. Houses built before 1980 may have lead-based paint or asbestos containing materials, which were outlawed in 1978 and 1981, respectively. Because most deconstruction projects will be on buildings older than this, it is likely that you'll encounter one or both of these materials (see chapter 5 for information on dealing with hazardous materials).

## Your Level of Involvement

How complex a project are you willing to undertake? A salvaging project can range from soft-stripping to whole-building deconstruction. Soft-stripping is the most common form of materials salvage because it is intended to recover only the components and materials that are most readily accessible, easiest to remove, and yield the highest value.

Soft-stripping typically doesn't disturb the structural integrity of a building, so it tends to be less hazardous in terms of physical safety. It can disturb various environmental hazards, however. Whole-building deconstruction is more comprehensive and involves the removal of all building materials, including structural components.

*Soft-stripping focuses on the more accessible items, such as interior trim, windows, and doors.*

*A full deconstruction is more comprehensive than soft-stripping and typically involves the salvage of the house framing.*

## Soft-Stripping

*Soft-stripping* is a commonly used term to describe the removal of nonstructural and easily accessible materials, typically components such as windows, doors, cabinetry, plumbing and electrical fixtures, and sometimes interior finish materials such as flooring, wall paneling, and ceilings. Soft-stripping stops at the surface. A full deconstruction requires a more serious level of materials removal, going beyond soft-stripping to include the roof, walls, and floor structures.

The decision to attempt soft-stripping or a full-blown deconstruction is based on the time available to do the work, the value and ease of transferring the recoverable materials, the scale of the building (industrial-size timber-frame warehouse vs. light-frame single-family house), available skill, and labor costs. Another important decision—and one that comes with a certain amount of risk—is how much of the building you decide to remove. If you plan to do soft-stripping you won't have to worry about removing walls, slabs, and footings. However, if you've agreed to take down a whole building, you may need to consider the use of heavy machinery to remove foundations and parts of the building that have little or no salvage value.

If you're planning to take down someone else's building, decide early on how much you are willing (or able) to commit to the effort so that everyone involved knows what to expect. A clear and concise written agreement with the owner will avoid hard feelings and possible litigation down the road. Questions to ask owners include the following:

→ Will they allow you to do soft-stripping only?

→ Will they allow you to deconstruct only down to the slab or footings (so you can avoid the cost of foundation removal)?

→ Do they expect you to leave a clean, reseeded lot with all evidence of the building removed?

Another consideration is how to deal with any debris already in the building and waste that you might generate in the deconstruction process. Many abandoned buildings contain the previous owners' belongings, which might be machinery and excess manufacturing supplies in an industrial building or furniture,

*The existing materials you have to remove will vary with building type and condition. This industrial military building has electrical lighting, plumbing, and fire protection systems that must be removed before timber harvest can begin.*

old clothes, food, and garbage in a single-family home. In all cases, these materials need to be dealt with before salvage begins. Also, not every bit of material in a building is salvageable and some gets damaged during deconstruction, so disposal is an important consideration. Will you need a roll-off trash container (Dumpster® is one brand), or do you have a truck? Where will you store the debris? How much are tipping fees at your local landfill?

For any salvage project, questions like these need to be clarified before the wrecking bars are dusted off. To help, we've developed a strategy to determine if a building is worth the effort, and we suggest some questions to ask so you can quickly determine if the building is a thumbs-up or a thumbs-down. If, after the initial survey, the building passes muster, an important next step is to take an inventory of building materials and a closer look to help uncover any hidden complexities or problems.

Quantifying the types and amounts of materials is important for determining the overall salvage value, building condition, and level of effort that is justified. The inventory also includes assessing for hazardous materials. It's prudent to assume that every building has some hazardous materials and should be approached with caution. Asbestos and lead-based paint are the two most common hazards in residential housing, though other hazards can be found in commercial and industrial buildings (for example, mercury switches, polychlorinated biphenyl [PCB] light ballasts). If in doubt, contact an approved environmental engineering company or consultant to take samples and test suspected hazards before the materials are disturbed. If hazardous materials are encountered during the deconstruction process, you need to stop working and have the materials abated before continuing work. This is a job for licensed professionals.

*In a house, you might find the abandoned belongings of previous owners, tenants, or squatters. All this needs to be cleaned out before unbuilding can begin.*

## Last On, First Off

Typically, a deconstruction project should follow the logic of the original construction process, only in reverse. An easy-to-remember acronym is LOFO—last on, first off. This concept is very helpful for older buildings in which it is not always obvious how something might be connected or where there may be hidden structural components. Deconstruction is typically a methodical process of unlayering a building, so a certain amount of patience and cautious curiosity are good companions.

## Making Sure the Building Is Sound

It is also critical to evaluate the building from an engineering standpoint to understand the structure and potential safety issues in removing load-carrying members. If you are unfamiliar with basic building construction, seek the advice of an expert. To paraphrase, remove no material before its time. This suggests that you remove something only when you are sure that it's not supporting any other part of the building. Obviously, you don't want to collapse part or all of the building unexpectedly.

An evaluation of the building is also a way to help plan labor and equipment needs, the sequence and schedule of deconstruction, and which parts of the building you intend to save or dispose. It's important to think through beforehand what parts of the building are structural and nonstructural and whether a material is to be reused, recycled, disposed, or treated as hazardous waste.

## Permits and Code Requirements

As part of the planning stage, it's a good idea to check with your building department about code requirements and the permitting processes for renovation, demolition, and construction. In addition, you want to make sure you address the requirements for proper disconnection of water, gas, electricity, and sewer lines that affect the property. If you are doing work for someone else, you may need a license.

If you know the building is older than 45 years (the minimum age to qualify for the National Register of Historic Places) and/or is in an area of historic buildings (which might be a designated historic district), you should check with either the local government historic preservation officer or the local preservation organization before starting work. Taking this precaution will help ensure that you don't remove a building you shouldn't. Also, these organizations can provide information about the building and sometimes have records on renovations and even old photos and documents that can help you better understand the building.

*For deconstruction,* first remove the skin of the building to expose its bones. Removing the layers one at a time in the reverse order in which they were applied is usually the best approach.

*The size and scale* of the unbuilding project can vary tremendously. The deconstruction and salvage of the 600,000-sq.-ft. industrial military building shown above was a massive undertaking. Similarly, the warehouse building shown below requires big equipment and considerable expertise to tackle.

## Making a Visual Survey

The first step in determining if your candidate building is a gold mine or a white elephant is to make a visual survey. Like the process of buying a used car, you'll want to do a walk around and kick the tires. The point is to get a general overview of the size, scale, and complexity of the building; its structural condition; site and interior accessibility; the presence of obvious hazardous materials; and a rough idea of what's available for salvage. Each factor considered at this stage is independently important in determining the feasibility of salvaging the building. The intent is that after this initial survey you should know whether to move ahead or walk away.

After discussing the more important items to consider in a visual survey, we provide an example of an actual walkthrough of a 1930s home under consideration for deconstruction and our assessment of the property (see "Case Study: Survey of a Deconstruction Candidate," on p. 48).

## Building size and scale

When walking around the building, evaluate its overall size and scale. The length, width, and number of stories are important in figuring the amount of salvage materials available and economies of scale in removing them. A large two-story building will have a higher ratio of materials to building footprint than a one-story building. However, the height of the building and the roof slope will determine if added equipment and safety precautions are necessary.

Roof slope is an important consideration for worker safety. Roofs with a pitch greater than about 5/12 (a rise of 5 in. vertically for every 12 in. horizontally) are difficult to stand on, and precautions must be taken to keep tools (and people) from sliding off. Higher work areas may require temporary safety measures to prevent or catch falls. Check with your local building department. Guidelines for fall protection and fall restraint requirements are available from the U.S. Occupational Safety and Health Administration (www.osha.gov).

*Unbuilding* a single-family home is manageable for most people.

*The low slope* on this roof (3/12) makes for easy and safe roofing removal.

*The steep pitch* of this roof makes fall protection a must for deconstructing this barn.

The size of structural members, the interconnectedness of building components, the number of walls and corners, and the complexity of the roof all increase the level of skill, time, and planning required to safely dismantle the building. Is the building simply constructed with a rafter roof or are complex trusses, bracing, and connections involved? Complicated rooflines may look great architecturally, but they may yield only short pieces of lumber that are tedious to disassemble.

## Structural condition

Of the many external forces that can wreak havoc on buildings, water is probably the most insidious. Water leaks can easily go undetected and result in decay, termite and ant infestation, and microbial growth that make materials unusable and create an unsafe and unhealthy working environment (more on this in chapter 3). Holes in the roof are an obvious indication that the interior has gotten wet and that rot may be present. Water stains on interior surfaces are a clear sign of water damage.

Keep an eye out for sagging rooflines, bowed walls, buckled or decayed lumber, insect infestation, and other indications that the structure of the building has been compromised. Missing shingles, flashing, or other roof elements can also indicate a problem, though such leaks are not easily isolated because water may follow pathways to other parts of the building well beyond the site of the leak. The same is true of plumbing leaks. We can't overemphasize the problem of water damage, which leads to rotted structural members, which in turn result in unstable and unsafe conditions. Be careful where you step when you're in any building that's sustained water or fire damage.

*While attractive architecturally, the small gables and shed dormer of this house are not likely to produce long lengths of reusable lumber.*

*The sagging rooflines* of these barns indicate structural problems with the framing. Caution should be exercised when evaluating barns like these for unbuilding.

*The bowed-out* column on this porch indicates a serious settlement or decay problem. Be careful where you step!

*Missing roofing* should raise a red flag and may indicate rotted timbers in the structure of the building.

## Site access

There are two things related to the site that make deconstruction safer, quicker, and more cost-effective: (a) clear access to all sides of a building and (b) open space for storing and processing (such as denailing) materials. Buildings without good access will likely slow down the deconstruction process because you'll have to clear space or force workers and equipment to maneuver around obstacles. For example, a corner row house in an urban environment is a more accessible site than a midblock location. Working on a site with tight access makes it difficult to store and load recovered material, load and unload roll-off trash containers, and park cars. Tight sites may force you to rely less on machines and more on hand labor.

## Interior access

Although the exterior accessibility of the site is important, so too is interior access. This is especially true of older homes that were built with narrow hallways, steep stairs, and a series of small rooms. Not only do interior partitions impede the flow of workers and materials and restrict the use of some equipment, but these walls require you to spend lot of time removing unsalvageable interior finishes (such as lath and plaster or drywall). The opportunity for accidents in close spaces, the reduction in natural light and ventilation, and the inefficient movement of materials make interior accessibility an important consideration.

*The severe decay* in the main floor beam of this old bungalow indicates that the floor might be dangerous to walk on.

*This row of houses, which were soft-stripped before being demolished for redevelopment, illustrates a less-than-ideal site. On a busy one-way street with no parking and little space between each building, the movement of people and the loading of materials were a challenge.*

*The narrow hallways* and stairways often found in older homes can make unbuilding more difficult because there is restricted room in which to work and it can be hard to move materials out of the building.

## Hazardous materials

Lead-based paint (LBP) and asbestos are the two most common hazardous materials found in buildings. Establishing the age of the building will help determine the presence of these materials before the expense of testing. Any structure built before 1978 is likely to have LBP and asbestos-containing materials. According to the U.S. Department of Housing and Urban Development (HUD), 74 percent of privately owned homes built before 1980 (57 million homes) contain LBP at levels defined in HUD standards.

Disposable lead test sticks are available that indicate the presence of LBP (more on this in chapter 5). The presence of LBP may require additional disposal costs or may result in the inability to resell an item coated with the paint, depending on the state you live in. Also, if the paint is flaking, measures should be taken to minimize the chance of LBP coming off during deconstruction and ending up on the ground, in clothing, in trucks, and so on. Anything that flakes off must be cleaned up.

As you do the walkthrough, take note of materials that might contain asbestos. Asbestos was used in pipe insulation, siding, roofing, drywall tape, and mastic for adhesives and flashing as well as in flooring products, such as linoleum, and their adhesives. If the property owner has already completed an asbestos survey, the report should note materials that are friable (asbestos in a form that easily crumbles), nonfriable, or suspect. If no survey has been performed and the presence

of asbestos is suspected, resolve this issue with the owner before signing a contract or starting work (or have a clause in your contract releasing you of responsibility of remediation). The removal of nonfriable asbestos, which may be left in place in a standard demolition and debris disposal, can significantly raise deconstruction costs. You are better off having the owner remove any asbestos as part of the salvage or deconstruction agreement. However, you may want to coordinate this process if any good salvage items could be damaged in the remediation process.

Questioning the presence of hazardous materials from the get-go will help determine whether some amount of abatement, worker-training and -protection, and materials-handling considerations will be sufficiently offset by the value of the materials recovered.

## Salvage potential

During the initial walkthrough, make a rough estimate of the quantity and quality of materials that might be salvaged for each of the main building

*Care must be taken to protect not only the workers but also the environment from peeling lead-based paint. Tarps should be spread on the ground to catch paint flakes before this siding is removed.*

*This surplus military building is sided with Transite®, a cement/ asbestos product. Care must be taken during removal; if the siding (which in whole form is considered nonfriable) is broken, asbestos fibers can be released.*

assemblies or systems: roof, floors, interior partitions and finishes, and exterior walls. Don't forget the finish flooring, kitchen cabinets, and fixtures. Note that much of the best lumber salvage is in wood-floor systems.

*In a single-family home, the largest-dimension lumber is typically found in the floor system, such as the 2×10 floor joists shown here.*

In 30 years, Pete and his wife Robin have deconstructed about 60 buildings, all of which were destined for the bulldozer. With the materials salvaged, they've built four new, modern dwellings and several dozen outbuildings, including garages, pump houses, garden sheds, and guest cabins.

"Daddy explained to me that it didn't do any good to save that old stuff unless you used it to build something new."

## Fifty Years of Unbuilding

## Pete Hendricks

To find someone who has built their life more completely around unbuilding and rebuilding than Pete Hendricks would be extraordinary. Pete started taking buildings apart over 50 years ago, when he was just a boy in rural eastern North Carolina. "When I was six years old, my daddy had me help him take apart an army barracks. It had been a temporary building used during World War II, but Daddy explained to me that the materials in the building were good so we wanted to save what we could before it got bulldozed. So I was out there in the cold, six years old, pulling nails and picking up bricks while Daddy took the building apart."

"When we got the old barracks taken apart we took all the materials over to our backyard and built a barbeque shelter-storage building. Daddy explained to me that it didn't do any good to save that old stuff unless you used it to build something new."

Pete got serious about unbuilding and materials reuse when he bought a small chicken farm in Wake County, North Carolina, after getting out of the Marine Corps in 1974. "All the outbuildings needed to be repaired so I went looking for some old buildings to tear down. What I found was abandoned farmsteads that were falling down or about to be destroyed to make room for development. People started giving me old buildings if I would take them down and clean up the site. Word got around that I did a good job. People started offering me more buildings, so I sold my chicken operation. My wife, Robin, and I started doing deconstruction full time."

It took Pete and Robin a while to accumulate enough deconstructed material to start building. The first thing they built was a lumber storage shed, and soon Pete and Robin had accumulated materials from about 15 structures—enough that they subdivided a piece of the farm to create some lots and started building new houses. "There was a lot to learn about using old materials in new construction, but we developed a process, a set of procedures. We organized the flow of materials from the old houses we were taking apart to the new houses we were building. Some of the materials had to be remilled, some of the materials needed only cleaning to prepare them for reuse."

"I've never seen deconstruction as a process unto itself. It's always been a part of the larger idea of converting old buildings into new buildings. However, my motivation has changed over time. At first I was just looking to get good lumber out of old houses, it was the practical thing. But I became enamored by the wood . . . the old heart pine, the white oak cut from huge trees, the cypress, cedar, and poplar we found. And I became motivated to streamline the conversion process."

"Deconstruction is a very physical activity. It's hard work, dirty, and sometimes dangerous," Pete continues. "The whole process requires a tremendous amount of attention to detail. On the other hand, we've never worked for anybody else, we've been able to work mostly outside, and we've spent our lives doing something worthwhile. Though I'm 60 years old and a bit tired, I've saved one building for every year I've been alive."

In lumber-industry parlance, the designation 2×4 is referred to as the *nominal* dimension, whereas the *actual,* or *dressed,* dimension is 1½ in. by 3½ in. We usually refer to lumber by its nominal dimension because it's a lot easier to yell, "Hey Joe, toss me one of those 2×4s, wouldya?," instead of "Hey, I need another 1½×3½!"

*The house* under consideration for deconstruction was recently bought by the church next door, which needed to expand its parking area.

Information on the building's history can help date original construction and provide dates of any modifications. This information, typically available from the local tax assessor's office or historical society, can help you determine what materials to expect. For example, plywood came into common use after World War II. Before the war, sheathing or subfloor materials were likely solid wood (typically 1×4s, 1×6s, 1×8s, and so on). Older growth, higher quality, and longer-length lumber and millwork are more likely to be found in a building built in the 1920s and earlier, though old-growth timber was harvested extensively until the 1970s. Don't forget your tape measure. Older lumber can have dimensions larger than lumber produced today (for example, a 2×4 may actually be 2 in. by 4 in., not today's 1½ in. by 3½ in.), making direct reuse with new lumber a challenge.

## Case Study: Survey of a Deconstruction Candidate

So far in this chapter we've presented general guidelines on what to look for in a building that is a candidate for deconstruction. To put these recommendations to work, let's look at an actual house and evaluate it for its unbuilding potential. Although the survey of a single house can't illustrate every decision you'll have to make, this example should give you a feel for some of the concerns and issues we would consider when surveying a house for deconstruction.

The house under consideration is located in an older neighborhood in a small town in central Pennsylvania. It was recently bought by a church whose main chapel is located next door. The church was considering three options for the house: renovation for church use, demolition to provide additional parking, or deconstruction.

Our first glimpse of the house told us two things. First, it appeared clean and well maintained, a real plus. Second, it did not contain the architectural embellishments that are usually the icing on the cake of a deconstruction project. Although it is always nice to find ornate light fixtures, stained-glass windows, and other architectural goodies, not every house is so adorned. Some salvagers would immediately take a pass on this house, simply because it lacks these more valuable and easy-to-strip items. Despite this, there is a fair amount of salvageable material in the framing and finished flooring of the house.

A couple of hours walking around and through the house gave us a good idea of the deconstruction potential and the types and quantities of materials available. At the time of our inspection, the owners had not completely ruled out renovating the house for church use. This affected our inspection process, because we didn't have the freedom to tear holes in the walls or ceilings or to inspect the underlying framing.

## Building size and scale

Town records indicate that the house was built in the 1930s. It is a functional two-story house with a half-story renovated space in the attic. The footprint is a square 24 ft. by 24 ft., and the house rests on a concrete-block basement foundation. The main body of the roof has a pitch of 9/12 (measured using a tape), and the front dormer is more steeply pitched.

The house has a large porch in front and a small one-story 9-ft. by 12-ft. addition in back. The simple form of the house and the single-level roof are advantageous for deconstruction. The front porch has a very shallow pitched roof and could be used as a working platform for the exterior of the upper stories in front. The house has been re-sided with aluminum siding, which means that the original wood siding is full of nail holes and unlikely to be salvageable.

In addition, the living space in the attic, which is accessible by an added stairway, could provide a convenient work platform for removing roof rafters from the inside of the roof, minimizing the use of scaffolding and other exterior platforms. The basement is one large unfinished space that could provide a place to work when removing the first-floor joists and would make it easy to remove all the main electrical, plumbing, and heating systems. The first floor is elevated about 2 ft. above grade. Given the simple lines of the building and its age, the framing appears uncomplicated and is most likely balloon framing.

*The distance between the sidewalk and the front porch doesn't provide much working space, a factor that needs to be considered when moving materials out of the house.*

## Site accessibility

The house is located on a relatively busy residential street in front (north) with a sidewalk and some mature trees and shrubs on the lot. It also has a neighboring house to the east and an alley to the west adjoining the church (to the right in the photo at left). There is a one-car gravel parking space behind the house on the south side. The alley is the principal access to the church's ministry building and the parking for the house.

*The house* is on a small lot bordered by a residential street, an alley, and a neighboring house on the east side.

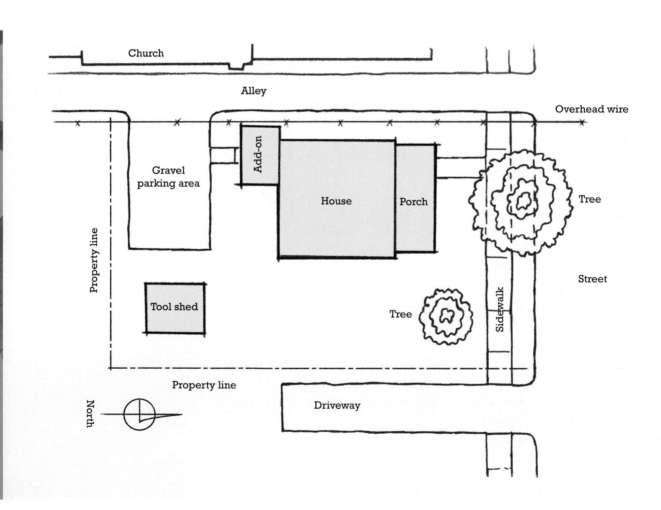

Church

Alley

Overhead wire

Gravel parking area

Add-on

House

Porch

Tree

Property line

Tool shed

Tree

Street

Sidewalk

Property line

North

Driveway

Several items need to be considered for site accessibility. Consider the front of the house first: The mature tree, the sidewalk, the street curb, and the shallowness of the yard all make this side of the house somewhat restricted for a roll-off trash container or for loading materials. A 20-yd. container set tight against the porch might fit but would block direct access to the front door of the house. The front doesn't look good for stacking and storing lumber for the duration of the project either because the material would be close to the street with increased risk of theft. In addition, because the lot slopes down from back to front, there are four steps up to the front porch (at the back of the house there are only two steps up to the door). Moving debris out of the front of the house would be problematic from the first floor, because there is not enough room to put a wheelbarrow ramp from the front door to the lip of the roll-off (the height of a 20-yd. roll-off is 4 ft. to 6 ft., depending on its length). So that rules out the front of the house for a trash container.

The alley on the west side of the house initially appears to be a good location for a roll-off, but the church regularly uses the road for deliveries and clergy access so we wouldn't want to block it. In addition, overhead electrical lines run along the alley just below the peak of the roof. A call to the roll-off hauler would be required to see how much clearance is needed for overhead access. (As a side note, we would also check with the electric utility company to determine if it requires that the wires be shielded during deconstruction because of their proximity to the house.)

Furthermore, there's only one small window on the second floor of the west side, which would make it difficult to drop debris directly into the trash container below. Finally, because the west end of the house is gabled, stripping the roof shingles would require building chutes to direct the debris into the container below. So the west side is ruled out.

*These overhead wires* may be too low to allow for unloading and pickup of a roll-off trash container on the west side of the house.

*The east side* of the house has enough room for a roll-off trash container, though the tree may need to come down or be trimmed. The five windows on this side of the house would make it easy to remove debris.

*The gravel parking* area behind the house provides extra working space, but the small addition adjoining it will need to be taken down first.

To the east, there's a side yard with a fair amount of room for a trash container. There are no overhead wires, though one tree might have to come down or be trimmed. Because that side of the house has larger windows on every floor, removal of debris out the openings would be very convenient. However, removal of roofing directly into a roll-off on this side of the house would be more difficult than from the front or the back. In addition, this side of the house adjoins the neighbors' driveway, so extra care would have to be taken to keep nails and other debris from being scattered into their yard.

So the east side of the house has potential for the trash container, but let's look at the back too (south). Here, there's a gravel parking area about 12 ft. by 20 ft., which seems a logical place for a denailing station and material storage. It might be a good idea to remove the back addition at the beginning of the project to create an even larger work area. Even better,

there is an existing shed in back that would make a good place for on-site storage of tools.

All things considered, the east side is the best place for the trash container. On this project we would want to use a 20-cu.-yd. roll-off, mainly because it's lower in height than a 40-yd. container and would fit under the first-floor windows more easily. A 20-yd. container is about 20 ft. long, 7 ft. to 8 ft. wide, and about 4 ft. high. Sizes vary by manufacturer so it's good to check what's available locally. Because the house is 24 ft. square, a 20-ft. container would nearly span the full side of the house, a real advantage in catching debris.

The back of the house doesn't appear to have enough room to store all the materials that will come out of the house, and the church doesn't have another storage area, so we would either schedule a pickup of a portion of the materials partway through the project or plan ahead for sales right at the site on a weekend.

*A trash container would fit on the east side of the house, or the area could be used for denailing.*

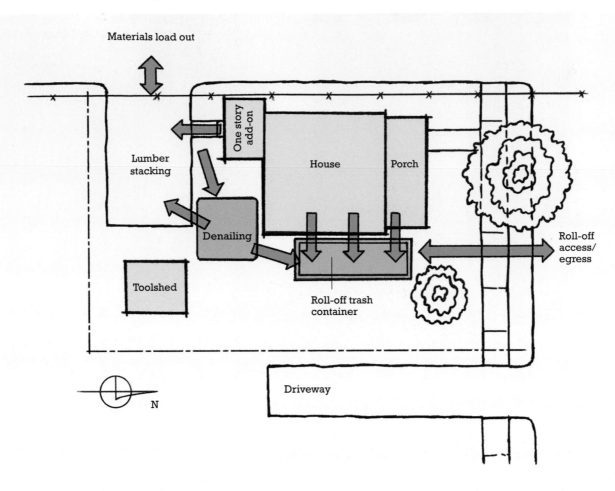

Materials load out

Lumber stacking

One story add-on

House

Porch

Denailing

Toolshed

Roll-off trash container

Roll-off access/ egress

Driveway

N

Rear add-on is removed to make extra space
between back driveway and house

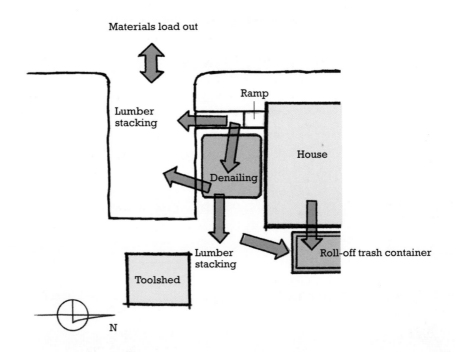

Materials load out

Ramp

Lumber stacking

Denailing

House

Lumber stacking

Toolshed

Roll-off trash container

N

## Interior accessibility

The first- and second-floor layouts are fairly typical of the era the house was built in—that is, a living room, dining room, and kitchen downstairs and three bedrooms and a small bath upstairs. The first floor is basically three large rooms, plus the rear addition. The second floor would be tighter to work in and might need to be opened up. The stairways are also narrow.

## Hazardous materials

Given the age of the house, we can assume that any wood surface under the aluminum siding has been painted with LBP. The single-glazed double-hung wood windows, which are flaking paint, appear to be original and are likely painted with LBP. We would also be suspicious of paint on the interior of the house and have it tested (for more on this, see chapter 5). Because the house has no spectacular architectural items, we assume we would have to dispose of these painted materials and plan on

using dust masks or respirators when we disturb anything with LBP.

There would be a couple of stages of concern for lead exposure with this house: when we take out the windows and doors and remove interior trim at the beginning of the project and then when we start removing the exterior aluminum and wood siding after the roof is off. There is some sheet linoleum in the kitchen and bath, though no apparent pipe insulation in the basement that might be asbestos based. There are no fluorescent lamps, but the thermostat looks old, so we will assume it contains mercury. We would include in the contract a condition that the owner perform an asbestos survey and remediate any found hazards before deconstruction work begins.

## Salvage potential

To assess salvage potential, we'll start on the outside. The house is covered with aluminum, a siding that is usually difficult to remove without

*The stairway is narrow, which can make removal of materials difficult and slow because only one person can move up or down at a time.*

*By peering under* the lowest piece of siding, we found that the aluminum siding sits over Styrofoam insulation applied over the original siding.

damage, so it will likely be separated out for recycling. When we look under the bottom row of siding, we can see a section of the exterior wall construction: The aluminum siding is covering a layer of ½-in. Styrofoam® board, which is over the original wood siding. Without pulling off some siding, it is hard to determine if there is board sheathing under the original wood siding. And because there is no firm decision whether this house will be renovated or torn down, we can't tear off a section of siding to get a closer look. However, the wood siding is flush with the exterior wall of the foundation, which usually indicates that there is no underlying

wood sheathing. With no wood sheathing and the wood siding likely full of holes, there isn't much salvage potential in the siding materials.

The kitchen cabinets are only average quality, and all the trim in the house has either been painted or is low quality. Although there are no hardwood floors, the softwood flooring under the carpets (likely Douglas fir) is salvageable if it comes up easily.

*Typical for a house* of this age, the original double-hung wood windows are covered with aluminum-frame storm windows for energy efficiency. The wood windows were in poor condition and not worth salvaging.

*The kitchen cabinets* are in good shape but not of the highest quality. All the trim in the house has been painted.

*The flooring* in the house is likely Douglas fir; if the quality found on the third floor indicates that of the rest of the house, it could be the most valuable material in the building.

Although the house is not rich with architectural items for salvage, the framing members offer better potential. To look at the framing, we start in the basement to verify the size of the first-floor joist framing and the flooring type. While there, we also check the condition of the chimney. In the basement, it's easy to see that the first-floor joists are 2×8s set 16 in. on-center. The joists are supported by a built-up beam (four 2×8s) at midspan. Because the joists overlap about 1 ft., each is 13 ft. long. All these joists are in good condition and salvageable.

*From the basement,* we can see that the first-floor joists are 2×8s set 16 in. on-center. The underside of the finish flooring is also visible, and it looks to be high quality and salvageable. There appears to be no subfloor.

*The chimney* is roughly centered in the house, so it will have to be taken down as each floor comes down.

*The mortar* of the chimney appears to be lime based (it powdered off freely when scratched with a key), which would make it easy to salvage the bricks.

*Look for plumbing* access panels behind bathroom fixtures. Peering inside can tell you a lot about the construction of the house.

There's no fireplace in the house, but a chimney runs from basement to roof to exhaust the water heater and furnace. Because the chimney is in the center of the house, we would have to take it down floor by floor, starting at the roof. Because of the age of the house, we would hope that the mortar is lime based, which would make it easy to salvage the bricks. Checking in the basement at the base of the chimney, we find that the mortar is a yellowish color (which indicates lime) and it powders off easily when scratched with a key. It should be no problem cleaning the mortar and salvaging these bricks.

Moving up to the first floor, we do not find many clues about the framing because we cannot look behind the walls and under the ceiling plaster. But up on the second floor we find a very helpful plumbing-access panel in the home's single bathroom. This provides us with a view into an interior wall cavity and a look at the framing and flooring of the second floor. A little measuring indicates that the second-floor joists, like the first-floor joists, are 2×8s set 16 in. on-center.

*The third-floor remodeled attic is very clean, with no garbage or other household items to be removed. The finished flooring is high quality and should be easy to salvage.*

We also find 1×4 tongue-and-groove wood flooring under the carpet but no subfloor. The interior wall framing appears to be standard 2×4s set 16 in. on-center. For making the preliminary estimates on recovery, we feel comfortable assuming these framing sizes and spacing are consistent throughout the house.

We next go to the third half-story under the slope of the roof that was made into livable space by building kneewalls and installing drywall. Fortunately, some of the kneewall paneling is missing, which allows us access to the roof framing. By crawling behind the kneewall, we're able to confirm that both the roof

rafters and the second-floor ceiling joists are standard-size 2×8s set 16 in. on-center. The roof sheathing is 1×6 tongue-and-groove. We also note that all the lumber found is of standard sizes—for example, the 2×8s are 1½ in. by 7¼ in.

## Is it worth it?

Based on our walkthrough, the framing and flooring appear to make up the lion's share of materials worth salvaging in this house. Using our field measurements, we were able to tally up the total amount of material salvageable. Accounting for losses, our tallies indicate that there are about 7,000 bd. ft. of 2×4s, 2×8s, and 1×6s in the walls, floors, and roof and about 1,800 sq. ft. of 1×4 tongue-and-groove Douglas fir flooring available. This represents a little over $4,000 in resale value, based on current used building materials prices. The chimney adds a few hundred dollars more in salvaged brick value.

If this house were in your area, would this be enough value to make the venture worthwhile? The answer to that depends on many other things in addition to the value of the materials. If you are paying a crew, how much are labor wages where you live? If you are doing the deconstruction yourself to reuse the materials, how much labor are you willing to expend for this value of materials? Is the owner expecting you to take down the house for the value of the materials or does he or she have a budget for removal that will pay some of your labor costs? An otherwise plain-looking house may look much more attractive, depending on the particulars of your project.

*From behind the kneewalls we could see the rafters of the house and verify that there were no subfloors.*

# THE MATERIALS YOU FIND

## Chapter 3

*The laborer is worthy of his reward.*
—1 Timothy 5:18

The biggest reward for your unbuilding efforts is the load of salvaged architectural items or the neat stack of building materials you will harvest. But because deconstruction is a methodical process of unlayering a building, you will have to remove, sort, and haul away a wide range of materials to end up with the mother lode of reusables you are looking for.

In this chapter, we discuss the different classes of materials you'll encounter (reusables, recyclables, and hazardous and nonhazardous waste) and your options for dealing with them. We also include a section on factors that affect the condition of building materials and signs you should look for to ensure that you don't waste time harvesting materials you can't use. Finally, we offer some thoughts on selling used building materials.

*Old-style claw-footed bathtubs are popular items to salvage, but they are heavy and require more than one strong back to move.*

Whether you are soft-stripping or a doing full-blown deconstruction, you'll be responsible for the safe removal, storage, and transfer of a variety of materials. Before you dive headfirst into a building, it's important to consider how you will remove the materials, what you will do with them, and if you'll need to store them. Tables at the back of the book will help you estimate the weights and volumes of commonly found materials (see p. 240).

All salvaged materials fall into one of four basic categories:

→ Reusable

→ Recyclable

→ Nonhazardous waste

→ Hazardous waste

### Reusables

We'll talk about the materials you salvage at much greater length later in this chapter, but an important point to consider up front is where you're going to store these materials. On-site storage should be considered as temporary, only until your deconstruction project is complete. Although longer storage might be negotiated, security and liability may be issues.

Storing materials outdoors is always a risky proposition. Nothing ruins your day faster than finding your neat denailed pile of 2×10s wet, warping, and moldy after a couple of cycles of sun and rain. Tarps help but aren't the best long-term answer. It's always better if you can get your materials under more permanent cover, such as in a secured garage, shed, barn, or warehouse.

Think about where you will store your materials offsite. Do you have a large enough space, such as a warehouse or barn that you can easily move materials in and out of? If you live in a one-bedroom apartment on the third floor, you probably don't have the space or access for 200 ft. of oak trim and a set of kitchen cabinets. And it doesn't take too many months of paying a rented storage locker fee to erase the value of your salvaged materials.

*Outdoor storage is risky in many areas of the country. This pile of 2×10s saw too much rain and started to grow mildew and mold, ruining their salvage value.*

*Tarps are only a temporary solution and will eventually leak or blow off.*

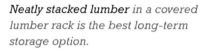

*Neatly stacked lumber in a covered lumber rack is the best long-term storage option.*

## Recyclables

A call to your city or county recycling department should give you the options for recycling and disposal where you live. The most valuable recyclable from deconstruction is the metal from plumbing pipes, heating ducts, metal roofing, gutters, siding, appliances, and so forth; most communities of size will have a company that recycles metal. It's worth the effort to separate metals into ferrous (steel and iron) and nonferrous (brass, bronze, copper, aluminum) piles because the nonferrous metal is more valuable. Copper, brass, and bronze are at record-high prices today. Although you won't get rich from the metals found in a typical house, you should make enough to pay for the gas you used to take it to the recycler and you won't have to pay to put it in the landfill.

*Your local **scrap metal dealer** will be interested in the scrap metals you generate and will likely pay you for them.*

**Recycling Appliances**

Appliances, such as refrigerators, can contain ozone-depleting gases, such as freon, and require special disposal. As a result, there is usually a charge to recycle appliances, depending on the type.

Finally, if you recover concrete, block, or unusable brick, there are other options than the landfill. Concrete recyclers are more and more common; and if one exists in your area, you might be able to drop off clean concrete for free (you may need to separate it from brick, and it may need to be free of lead-based paint, so call first). Another option might be to drop it off as fill material. Earthwork construction projects often need clean fill for filling in an excavated area. Concrete, block, and brick qualify, as does clean dirt. No wood, garbage, or other organic materials are allowed. Frequently, you'll see a handwritten sign near an earthwork project that reads, "Clean Fill Wanted" or you may find a want ad in the classified section of the local newspaper. Call first, but you can often drop off appropriate materials for free at these sites.

*Many communities have a recycling yard that will take household metals for free, though there is usually a charge for appliances.*

The recycling of other building materials varies by location. Drywall can be recycled in some communities, though it is usually limited to clean, unpainted material. Clean scrap wood is often recycled into mulch, but pressure-treated wood is not allowed and nails may or may not be tolerated in the grinding machinery. Ask the recycler. Businesses have sprung up to recycle asphalt shingles (nonasbestos only); others recycle carpet and padding. Yard waste such as brush and small trees may require separate collection and processing at your city or county recycling center. Again, your recycling coordinator should be able to help determine local options and requirements.

*Wood recyclers will accept clean wood scrap for recycling, but the pressure-treated wood shown here (green) would not be allowed. Treated wood should be disposed of in a construction and demolition (C&D) landfill.*

## Nonhazardous waste

By reusing or recycling the materials you generate, you should be able to keep the largest portion of the building volume out of the landfill. However, it's inevitable that you'll generate some waste: materials get broken; commingled; or simply can't be reused, recycled, or given away. The only other option is the landfill.

Many communities have two types of landfills: one for traditional household garbage collection (municipal solid waste, or MSW) and one for construction and demolition (C&D) waste. Most of what you'll generate in deconstruction is appropriate for a C&D landfill. Typically, C&D landfills operate under different rules and have different costs (or tipping fees) than MSW landfills. It is important that you keep the household and C&D waste materials separate because C&D landfills can reject loads containing household waste or charge a higher rate.

## Hazardous waste

You're likely to find several types of potentially hazardous materials on an unbuilding project, including asbestos, lead, mercury, polychlorinated biphenyls (PCBs), unlabeled containers of mysterious substances that might be solvents or oils, and other chemicals in older buildings. We talk in more detail about these materials in chapter 5, but here we offer some quick thoughts on dealing with some of them and the common places to find them.

Some of the most obvious hazards in a home are the chemicals left behind by previous owners. Household chemicals might be under the sink or in cabinets, and pesticides, poisons, oils, paints, and preservatives might be hiding on a shelf in the garage, shed, or basement. If there's a local household hazardous waste facility in your area, you'll likely be able to dispose of these items for free. Otherwise, you'll need to plan for proper disposal. Your county recycling coordinator should be able to help. Old refrigerants (some containing chlorofluorocarbons; CFCs) in heating and air-conditioning equipment could also be present, and you will want to take care that none is released into the environment.

Lead is usually found in the form of lead-based paint (LBP), but it can also be present in metallic form—for example, as a sealer in cast-iron waste lines and as sheeting in roof-vent pipes. Metal lead can be recycled with your local metal recycler; however, the proper disposal or

*Most cities have no-cost collection facilities for household chemicals and paints that ensure proper disposal. These facilities often include a material exchange to encourage the reuse of more benign chemical products (paints, cleansers, and so forth).*

*Fluorescent light tubes may require special disposal. Check your local requirements before throwing them in the trash container.*

*The black box inside a fluorescent light fixture (the "ballast") can contain PCBs if the fixture is old.*

allowable reuse of materials painted with LBP varies, depending on the state where you live. Mercury and lead can be found in fluorescent light bulbs, high-intensity lights, old light switches, thermostats, and old thermometers. Federal law and some state laws restrict the disposal of used fluorescent light tubes and high-intensity discharge (HID) lamps, including mercury vapor, high-pressure sodium, and metal halide lamps because of the presence of these heavy metals. Fluorescent tube recycling is becoming more common, so look for this service in your community if you have to deal with these lights.

PCBs can be found in older fluorescent light fixtures. The ballasts in these older fixtures should be taken to your local household hazardous waste facility. Generally, if a ballast is electronic or is labeled "no PCBs," it can be thrown in the trash. If, however, it contains oil and is not labeled "no PCBs," it must be disposed of as hazardous waste (often at the household hazardous waste facility). Make a point of finding out local rules for disposal or recycling of each type of material. Ask about which materials are accepted; the condition and sizes of material; and, of course, the cost and hours (or season) of operation.

## Assessing What's Reusable

To help you get the most out of the building you deconstruct, in this section we share our experience with the materials you will find and provide some thoughts on reuse potential, material types, and material condition issues. In addition, we offer some tips on calculating quantities of common materials so you can make realistic estimates of yield, which will affect your labor input and overall salvage return.

All else being equal, the quality of the materials is more important than the quantity. Unique or historic architectural components, such as stained glass, fireplace mantels, ornate doorknobs, and other fancy hardware, always command the

highest prices and require relatively little labor to remove. Harvesting the cabinets, plumbing fixtures, windows, doors, siding, and lumber requires more work. For this reason, we'll first discuss the materials generally removed during soft-stripping and then move on to materials typically removed in a whole-house deconstruction.

Finally, we review the measurement of materials for inventory purposes. Although it's important to quantify the materials in your building, you don't have to go overboard measuring, especially with materials that will be disposed of or recycled. In other words, you don't need to measure the building dimension to the nearest inch to

*Salvaged lighting can vary tremendously in quality and style, depending on the age of the house and the taste of the owner.*

determine the amount of roofing you will remove. However, you don't want to measure too roughly either because the dimensions of some materials, like lumber or cabinets, need to be known more precisely.

## Electrical fixtures

Lighting is one of the easiest and quickest items to remove from a house, generally requiring only a ladder, a screwdriver, wirecutters, and pliers. Although newer lighting has some value, it is generally the vintage fixtures that are worth the most. There is a ready market for vintage lighting for historic and period home restoration projects. Of course, the style, age, uniqueness, and condition of the lighting greatly affect its value.

Usually, the older the fixture, the better. Bronze or brass light fixtures are generally more desirable than ferrous metals, which were often painted.

As with any antique, lighting that has its original finish (and original glass shades) is the most desirable. Completeness is also important: missing glass, chains, or other decorative elements are difficult to match or track down. For example, a missing or damaged original glass shade on a four-light ceiling fixture reduces its value by at least half. Also, most iron or pot metal fixtures were decoratively painted. If ceiling paint was slopped on the fixture over the years (or worse yet, the whole fixture was painted), you'll find it difficult to remove the ceiling paint without removing the original decorative paint (if the paint is latex, soaking the disassembled fixture in hot water can soften the paint).

Painted bronze, brass, or copper lighting is more forgiving because the paint can be stripped and the fixture buffed to a nice finish. Take care when removing lighting that you keep fasteners, ceramic insulators, and other parts together with the fixture. A zip-closing plastic bag comes in handy here; fill it with the small parts and tape or tie it to the *back* of the fixture.

Other vintage electrical items worth keeping include brass, bronze, and decorative switch plate and outlet covers.

*Vintage electrical lighting is highly desirable, but condition is very important. These almost pristine ceiling fixtures are readily salable.*

*This salvager* saved the cost of a new heating system by removing these working furnaces and their associated ductwork from a deconstructed home.

## Heating and air-conditioning equipment and other appliances

While furnaces and hot-water heaters are not as exciting a find as an original stained-glass window, there is value in this equipment, depending on its age and condition. Generally, the newer the better. (Many Habitat for Humanity ReStores do not accept any equipment if it's older than 5 years.) Ideally, furnaces should be high efficiency and water heaters in excellent shape and less than a few years old.

Air-conditioning units and fans can be salvaged and reused, but value varies with local market demand. A whole-house unit will require disconnection by a heating and air-conditioning technician to ensure that no CFCs are released. Wood stoves and gas fireplaces require little work to disconnect but a strong back to move from the building. Keep an eye out for stainless-steel stovepipe. It is very expensive and worth saving when salvaging a wood stove. Old steam or hot-water radiators, if decorative, can be resold for historic restoration projects.

*Many used building* materials stores keep furnaces, water heaters, and wood stoves in stock.

*Decorative vintage* hot-water or steam radiators can be resold for historic restoration projects.

The practicality of salvaging standard kitchen appliances, such as refrigerators, stoves, dishwashers, garbage disposals, and ovens, depends very much on condition, age, and brand. Electric appliances are more salable if they're electronically controlled (i.e., electronic push button vs. mechanical dial). Many buyers (especially landlords looking for replacement appliances) are less fussy about the age of gas appliances since a 1970s gas stove operates pretty much the same as a 2007 model. Condition, color, and cleanliness, however, are important. White and stainless steel seem to be perennially popular. If you're lucky enough to find high-end appliances that haven't been abused, they are worth snagging, especially if they are restaurant grade.

*This vintage stove salvaged from a 1920s home was in mint condition and had correspondingly high value.*

*Homeowners expressed definite interest in the electrical panels offered at this used building materials auction.*

*The value of traditional-styled plumbing fixtures depends on local markets.*

## Electrical equipment

For the most part, the electrical system in a home is not reusable, though the wiring can be recycled along with any metal conduit that is found. Older electrical switches aren't worth much, though some used building materials stores carry them. For someone doing a historic preservation, the two-button light switches found in some old homes may have value. Ungrounded two-prong outlets are generally not worth keeping. If you deconstruct a house with a newer main electrical panel (or subpanel), the box and breakers may be worth saving for resale to a DIYer. Many electrical contractors will not install used electrical boxes or breakers for liability reasons.

## Plumbing fixtures

Toilets, tubs, and sinks probably aren't the biggest sellers in a used building materials store, though many carry an inventory to help customers who might need to match an older color or style. Like everything else, the value of these items depends on age, condition, and brand. Water-conservation laws in some states have outlawed the use of old high-flow toilets, which has reduced the value of many older commodes.

*Stainless-steel sinks never seem to go out of style and are usually easy to resell. Vintage pedestal sinks are also popular.*

*This original 1920s stilted sink salvaged from a California bungalow is a good find.*

Vintage clawfoot tubs and pedestal sinks are keepers in the reused plumbing fixtures market. As with lighting, condition is extremely important. An old tub or sink that is rust stained and chipped will not bring in top dollar because it will need costly refinishing. White is the most desirable color, and longer clawfoot tubs (especially 6-footers) are more valuable than shorter ones. The style of the feet on clawfoot tubs plays a big part in determining their value; generally, the more ornate, the higher the value. If the tub or sink has original porcelain faucets, they may be worth saving and selling with the fixture.

Cast-iron built-in tubs and wall sinks are less desirable, unless they are very high quality, and often end up as scrap due to the hassle of moving them. Fiberglass shower units and whirlpool tubs can be resold, but need to be removed carefully to prevent flexing that may crack the unit.

*Clawfoot tubs are expensive to refinish, so a clean tub with no chips or rust stains is the best find. Be sure to keep the feet matched with the tub or the value will be reduced.*

**Removing Cast-Iron Fixtures**

If you find that it's not worthwhile to remove an old cast-iron fixture (tub, sink, and so on) whole, the brittleness of cast iron makes it easy to break into manageable pieces with a sledgehammer. Make sure you wear gloves and eye protection because the porcelain coating on the fixture shatters into sharp shards.

*Kitchen cabinets are usually easy to remove from a deconstruction project. These cabinets are ready to reinstall in a new space.*

## Cabinetry, casework, and stairs

Cabinetry, especially kitchen cabinetry, is usually one of the easiest and first things to harvest from a deconstruction project. The desirability of a particular type of cabinet varies, depending on what it's made of, how well it's put together, and how it's fastened to the wall. Also, cabinetry styles come and go, so some cabinets are simply out of fashion for reuse. Even so, if the cabinets are reasonably priced, you may find a market for people looking for storage space for workspaces, garages, and outbuildings.

Cabinets can be awkward to handle and take up a lot of space; and because units need to match the design of the new location, they are less flexible to market. On the other hand, a single-piece upper cabinet is easy to remove and sell as a single self-supporting unit. Modern modular cabinets are usually more marketable because the sizes are standardized and the units are easy to reinstall. Built-in cabinets can be difficult to remove since there always seem to be a lot of nails holding each one in; and once the cabinets are removed, they can turn out to be odd shapes and sizes that are difficult to integrate into a new space.

*These original 1920s cabinets were in great shape and, though built-in, were well worth salvaging and reusing.*

## Kitchen Cabinet Sizes

Modern kitchen cabinet sizes are standardized. Base cabinets are usually 24 in. deep and upper units 12 in. to 14 in. deep. Widths vary depending on the function of the cabinet but typically range from 9 in. to 48 in., in 3-in. increments. All else being equal, a kitchen full of various-width cabinets is relatively easy to reinstall because of the flexibility for fitting the cabinets into a new space.

Cataloging cabinets not only helps in inventory for resale but also helps a customer plan for reuse. For example, the data in the following list should provide all the information necessary for a buyer to plan a project:

- Oak cabinets (solid fronts, plywood boxes)

- 1 each, 32-in. lower, double door, 3 adjustable shelves, finished left side only

- 2 each, 18-in. upper, left-swing door, 2 fixed shelves

- 1 each, 36-in. lower, sink base, double door

In addition, a photo of the cabinets in place before removal can help a buyer visualize how they might work in a new project.

Cabinets are like furniture, and solid wood is more desirable than fiberboard, plywood, or particleboard. Most sought-after are hardwoods such as oak, cherry, hickory, and walnut, though each seems to be in vogue at different times.

Don't pass up opportunities for soft-stripping cabinets from schools, hospitals, and office buildings when they remodel or rebuild. Although

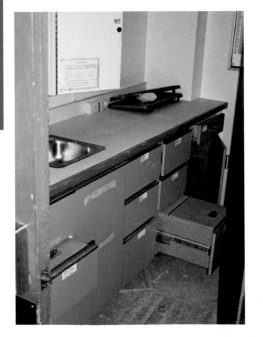

many items from these buildings are commercial size and impractical to use in house remodeling, such as 42-in.- and 48-in.-wide full-height doors, cabinets are generally the same size as home cabinets and can easily be removed. Individual elements like mantelpieces and stairs can range from very high value (usually if ornate or period correct) to not worth the effort. Stairs are often difficult to market because they can be difficult to fit into existing construction and generally can't be altered without considerable effort.

*Remodeled or razed public institutions such as hospitals and schools can be a source of usable well-built cabinets. The cabinets shown in the photo at left are from a medical clinic, and those in the photo above are from a university laboratory. Be aware that older slate-like laboratory countertops can contain asbestos and should not be cut.*

*Most contemporary wood trim is salable, though not of high value.*

*Period fireplace mantels and surrounds are highly desirable, but they require careful removal to prevent damage.*

*Period wood trim, such as this original-finish oak baseboard, has more value than painted trim.*

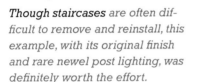

*Though staircases* are often difficult to remove and reinstall, this example, with its original finish and rare newel post lighting, was definitely worth the effort.

*This elegant staircase* was salvaged from a high-end teardown property.

*If the whole staircase* is not worth removing, salvaging stair parts like these newel posts and balusters can be worthwhile.

## Doors, windows, and shutters

Doors themselves are easy to remove and, as with everything else, quality and condition are paramount. A painted hollow-core door is usually not worth much; but an original-finish, Victorian solid-oak entry door with beveled glass, dentil molding, and original hardware would be a real find. Interior solid-core doors with individual wood or glass panels (French doors) are very salable if they are standard width (32 in.) and height (80 in.). Exterior doors are typically 36 in. wide and heavier duty.

Reused doors are usually bought for one of two purposes: to rehang the door as a replacement in an existing opening or to hang a door *and* frame in a new rough opening. In our opinion, for any door that is vintage with matching finish on the door and casing (without paint), it's best to remove the finished frame of the door, keeping all the parts together so the door can be more easily refit in a new rough opening (more on this in chapter 6). Bundle together all the trim pieces and hardware (screws, pins, and other small parts can go in a zip-closing plastic bag) and keep them matched with the door. It's a good idea to number the trim pieces so the reinstaller knows how they fit back together. Write the number on the back of the piece.

Windows are usually a little more difficult to remove than doors because one side is on the exterior of the house and removal can require working at height. So, to remove them safely, you will need a ladder or work platform to access the exterior trim. Older single-paned windows, which inevitably will be painted with LBP, may not be worth saving, but unpainted true divided-light windows can have value for restoration or craft projects. Double-glazed windows are well worth saving, but you are less likely to find them unless you are taking down a new or recently renovated building.

*Solid-wood paneled doors are much more salable than hollow-core doors.*

*It's not unusual to find fancy hardware in homes slated for demolition.*

*Original hardware can enhance the value of a recovered door or can be sold separately. Glass knobs are generally the most valuable (colored even more so), brass next, and steel the least desirable. Porcelain knobs also have value. Try to keep the doors, knobs, and hardware as a matched set when possible.*

Don't forget about shutters. Period shutters can be used in historic restoration and other decorative uses if they're in good shape. Of course, finding stained-glass windows in a house is always a treat because they have high market value, depending on design and condition.

*Most windows are relatively easy to remove from the inside, but you may need a ladder or platform for safety on the exterior.*

*Energy-efficient windows can be salvaged and easily incorporated into new construction.*

*These stained-glass windows were salvaged from vintage homes and churches slated for demolition. While stained glass is always a good find, you have to be very careful when removing to prevent breakage, which is costly to repair.*

*Removing the base shoe and inspecting the flooring will indicate if the floor has been sanded. The photo at right shows a ridge where the floor sander stopped, indicating that this floor had been sanded at least once before.*

## Tile, vinyl, and carpet flooring

Ceramic or porcelain tile is very difficult to remove from a floor without destroying it and is generally not worth the effort unless the tile is one of a kind or historic. Vinyl flooring is typically glued down and not salvageable. Resilient flooring materials, such as sheet or tile flooring, are typically laid over an underlayment (often $1/4$-in. plywood), so when assessing what a floor is made of, dig deep enough to understand the composition of the floor. Sometimes you will find vinyl laid directly over a wood floor, which can make removal and resanding the flooring difficult. Carpet is typically not salvaged, unless it is nearly new and in large pieces. It does however make good padding for stacking materials for moving, and it is also useful to lay down under the denailing station to catch nails.

*Carpet can be resold, but only if it is very clean.*

*Sometimes you have to dig* for the treasure. *This original oak staircase was hidden by a wall of drywall.*

*Acoustical tile* *can be recycled as long as it doesn't contain asbestos.*

*Tin ceilings* *can be removed, but it takes a lot of care to prevent bending and denting. Expect some losses, especially in corners and along the edges of the walls.*

## Ceilings and interior walls

In most cases, the materials used for ceilings and walls are either drywall or lath and plaster, which aren't reusable; so these materials are overburden you have to remove. You may find wood paneling in various patterns, such as bead, V-groove, and raised panel. If solid wood and unpainted, it can have good value and is relatively easy to remove. Hardwoods such as walnut, cherry, or mahogany would make the paneling even more valuable. Take care in removing wood paneling because it can be dry and brittle and chip or split, which will reduce its value. You should always try to determine if additional layers of interior finishes have been added that could be hiding something great, such as original wood paneling, hardware, or other treasures.

If you are lucky enough to find a pressed tin ceiling in good shape, be aware that it is probably the most valuable ceiling finish found in old buildings. It needs to be removed carefully because dents and creases lower its value. Acoustical tile is common in many post–World War II homes and commercial buildings and was typically applied either to furring nailed to an existing ceiling or directly to the drywall or plaster. Depending on ceiling height, you may encounter a dropped ceiling, which is either framed down with 2×4s and finished with drywall or acoustic tile set into a suspended metal grid. If you find a lot of unpainted acoustical tile, it may be recyclable if it doesn't contain asbestos.

## Insulation

You'll find three basic types of insulation: loose fill, batts, and rigid foam. Loose fill, which includes blown cellulose or fiberglass and poured vermiculite, is messy and difficult to collect and usually isn't reused. However, if you have a use for it, you can fill garbage bags with material collected from the walls and ceilings. Batts, usually fiberglass (though sometimes mineral wool) can be salvaged and reused if clean and in good shape. Rigid foam insulation is worth saving if the pieces are big. Don't try to reuse any insulation if it is wet and/or moldy.

*Loose-fill insulation, such as the vermiculite shown here, typically isn't worth salvaging. However, if you have a use for it and you have the time and patience, you can scrape it out of the walls and ceilings and reuse it. If you find vermiculite insulation in an older home, have it tested for asbestos.*

*This 3-in.-thick rigid insulation is well worth saving.*

*Batt insulation can be salvaged and reused if carefully removed and if it is in good shape. Roll it up and place in plastic bags for transport.*

*This salvaged 8-in.-wide horizontal cedar siding is painted on one side; but it is of very high quality, as indicated by the clear back side, and can easily be reused.*

## Exterior siding: wood, brick, vinyl, stucco

Siding can vary tremendously in quality and condition. It is wise to determine if there's a single layer or multiple layers of siding as more layers make it all harder to remove. Also, inner layers can be damaged from the nails used to apply the outer layer.

### WOOD

Wood siding comes in many species and sizes, though typically naturally durable species, such as cedar, redwood, and cypress, were used because of their superior performance in exterior exposure. Each building has to be evaluated individually because—frankly—depending on the age of the house, you never know what you'll find. The difference can mean a material not worth the effort to save (such as log-grade yellow pine) to a material very worthwhile saving (clear-grain Douglas fir, redwood, cypress, or cedar). As with interior wood paneling, exterior siding takes great care to remove without damage. Because siding was one of the highest-grade wood products produced, long clear (that is, no knots) lengths can be found. If the siding is removed intact, it is relatively easy to reuse in another project.

## BRICK

Brick salvage and reuse depends on two things: the quality of the brick and the ease of mortar removal. Portland cement–based mortar, which has a white or gray appearance, sticks tenaciously to brick and makes cleaning difficult. Older lime-based mortars, which typically have a distinctive yellowish color, are much easier to remove and are found in older buildings. If the mortar is soft and comes off in a fine powder, it is lime based. Bricks from pre–World War II buildings may very well have lime-based mortar.

Bricks require a lot of work but if the quality is high and they are set with lime-based mortar, they may be worth it. Think about it up front, though, because bricks are heavy. It usually takes a lot of bricks from a single source for them to be salable, and a lot of handling is required to strip mortar and to stack. You will also need to use pallets and a lift of some kind to move the brick from the site. Extruded bricks with holes have

*Brick construction is very common in many inner cities; the type of brick and mortar used will determine salvage potential.*

**Sidewalk Salvage**

Some of the most desirable bricks are those that were used for street paving, because they're larger and more durable than those used in houses. If you see an old brick street being taken up, the bricks are definitely worth saving if you can make a deal to obtain them.

very limited options for reuse; and if they're bonded with cementitious mortar, they're generally not worth trying to recover for reuse. Brick rubble can be recycled for fill material. Check with your local concrete recycler to see if they will take brick and masonry along with concrete and the conditions or degree of separation they require.

## Recovering Landscaping Materials

While building materials are the obvious target in a deconstruction project, landscape materials such as paving stones, brick, and natural stone garden borders can often be easily removed (if not cemented in) and reused. Fences, arbors, and gazebos are sometimes available. Also, depending on species, ornamental shrubs and perennials can be dug up and resold. We know of at least one organization that specializes in these recovered materials and saves hundreds of ornamentals each year for resale and reuse. Make sure you have a clear understanding with the owner before taking these items.

*Shrinkwrap palletized bricks to hold the stack together.*

*These landscaping pavers are as good as new and well worth salvaging.*

## VINYL

Vinyl is sold at more and more reuse stores, and it is becoming more common to see it reused. As with other types of siding, it's important to remove it in the correct sequence. The first piece has a flange that locks it to the overlapping next piece. Start from the top to reveal the fastening edge to remove the nails and allow the upper piece to be pulled off the piece below. If you break the nailing flange or the lower locking flange, the piece is unusable.

## STUCCO

Stucco is a durable and long-lasting siding material that's difficult to remove and not salvageable. Factor in extra labor if your candidate building has stucco and you want to do a full deconstruction. Because traditional stucco was typically built up in three coats (with the first coat applied over a wire mesh), removal usually involves a lot of heavy hammering to break it into manageable pieces. We would approach the full deconstruction of a stucco home with some apprehension.

## Roofs

If you're lucky enough to find a building with a slate roof or clay roof tiles, these can be very worthwhile, though laborious, to salvage. Both tile and slate vary in quality. Some slate is hard and tough without being brittle. Soft slate is more subject to

*If you have enough pieces of the same color and style, vinyl siding can be resold. It can also be stored outside.*

## Estimating Roofing Materials

For roofing and roof sheathing a little trigonometry helps when calculating roof area and estimating quantities of roofing materials. Recall that the length of the hypotenuse (long side) of a triangle squared equals the sum of the squared lengths of the two right-angled sides ($A^2 = B^2 + C^2$). If you're measuring a gable roof and you know the width of the building, estimate the height from the top of the walls to the ridge by measuring at the gable end. This gives you the two legs of the triangle you need to calculate the length of the slope of the roof from the top of the wall to the ridge (see the drawing on the facing page).

erosion and to attack by airborne and rainwater-borne chemicals. Look for wear at the nail holes, delamination of the tile, and dusty flakes coming off the tile, which indicate that the tile is probably too soft for reuse. Tile and slate can both break if you're not careful walking on them, and care must be taken in removing the fasteners that secure the individual tiles to prevent damage. Various types of metal roofing, such as standing seam and metal tile, can also be found and, depending on condition, are worth salvaging (or recycling).

If you find a slate or tile roof to be salvaged, taking pictures of the roof before salvage will help a potential buyer visualize how the roofing will look on their project. We suggest you measure the roof area and convert it to the number of "squares" of roofing so a prospective buyer (or his or her roofer) knows how much area the roofing will cover (a roofing square is 10 ft. by 10 ft., or 100 sq. ft., of *installed* roofing).

*Salvaged slate* has high value, though care must be taken to prevent breakage during removal.

## Calculating Gable End Wall Area

1. Multiply width by height
2. Multiply *A* by *B*
3. Add the results of 1 and 2

## Calculating Sidewall Area

Multiply height times length

## Calculating Roof Area

1. Measure *A*
2. Measure *B*
3. Calculate *C*, which equals $\sqrt{A^2 + B^2}$
4. Add overhang width to *C* and call the result *D*
5. Multiply *D* by the length of building; multiply that result by 2

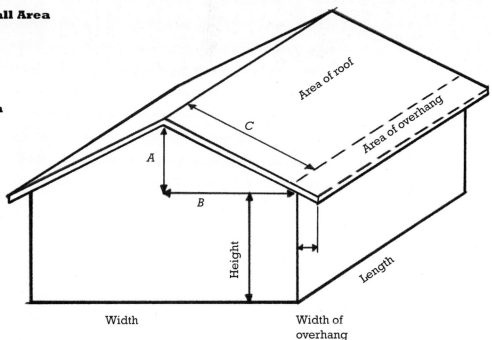

As you estimate the amount of roof sheathing that is salvageable, take note of the types of boards used and the number of layers of roofing. Typically, lumber used for solid-wood sheathing wasn't the highest grade and, given the damage due to nailing, can make the salvage of these boards questionable. Another potential problem is that in some roofs the building paper has melted and left a bituminous residue in the sheathing. If practical, try tearing off a section of roofing to see how the sheathing boards look.

*If not too rusted, tin roofs can be salvaged and reused.*

## Wood framing

It's often difficult to visualize the bones of a building with the wall, floor, and ceiling skin intact. To determine the sizes and volume of framing with these materials in place usually involves breaking a hole in the first-floor ceiling or crawling into the attic to measure the rafters and ceiling joists. First-floor joists should be accessible from a basement or crawl space. This will tell you not only the joist size (usually 2×6, 2×8, 2×10, or 2×12 in most homes) but also the joist spacing (usually 12 in., 16 in., or—more rarely—24 in. on center). The length of a joist is usually outside wall to outside wall in a simple structure, but if there are interior bearing walls, joists may be spliced over those walls, so you may need to break a hole in the ceiling near the bearing wall to determine joist length.

Roof rafters can be seen from the attic and first-floor joists from the basement or crawl space. Wall stud size is almost always 2×4, though sometimes 2×6 in new construction.

In any case, a look at the wall thickness at a doorway will tell you the stud size used. Determining the amount of lumber you have in a building is simply an exercise in counting up the pieces of a given size and length.

In our experience, a minimum useful length for 1× lumber is 3 ft. and for 2× lumber, 4 ft. Most Habitat for Humanity ReStores accept only 6 ft. and longer 2× lumber, but a 4-ft. length is useful for a variety of building purposes (such as cripple studs, window framing, headers, and blocking). Assuming a 25 percent loss for breakage during salvage and trimming to length is reasonable.

*These first-floor joists* overlap on a steel beam in the basement. Second-floor joists often do the same over a bearing wall, but it is not obvious unless you break a hole in the first-floor ceiling to verify.

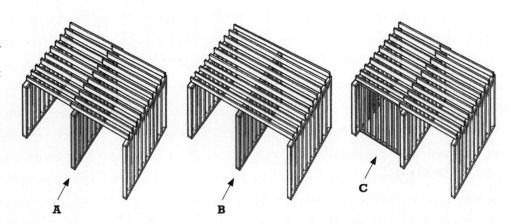

**KNOW YOUR BEARING WALLS**

Knowing if a wall bears a load will let you know if you can safely remove it. Wall A is definitely load bearing because the joists are spliced over it. Wall B could be a bearing wall if the joists don't span the full building width. Wall C is not load bearing because it is parallel to the joists overhead that carry load above it.

**A** Definitely a bearing wall

**B** Maybe a bearing wall

**C** Not a bearing wall

A     B     C

*Crawling into the attic* can tell you a lot about the anatomy of the roof and allow the close inspection of rafters, rafter ties (horizontal pieces in this photo), and ceiling joists.

*Wall studs* in houses are usually 2×4s spaced 16 in. on center.

## Figuring Board Footage

It's handy to be able to quickly calculate how many board feet there are in a stack of lumber. A board foot (bd. ft.) is a standard measure of lumber volume and is a board that's nominally 12 in. wide by 12 in. long by 1 in. thick. So to calculate the board footage of a particular board, you need only to multiply the *nominal width* (in inches, *w*) by its thickness (in inches, *t*) by its length (in feet, *L*) and divide the result by 144. The following formula will do it:

**(w × t × L × 12)/144**

For example, if you have a 20-ft.-long 2×12, it contains 40 bd. ft.:

**(12 × 2 × 20 × 12)/144 = 40**

If your board is a nonstandard size, you can calculate the *actual* board footage (same way as above, only use the actual width and depth dimensions of the board). Lumber calculators are available online (for example, try www.woodweb.com/Resources/RSCalculators.html). Let your customers know if your calculation is based on nominal or actual dimensions and price accordingly.

## MEASURING LUMBER

When estimating salvageable wood in your deconstruction project, make sure your estimates are as accurate as practical. On one hand, you don't want to measure with more precision than necessary; yet on the other hand you don't want to be so loose that you can't describe the material properly to someone who might want to buy it. You also should be clear whether you are talking about actual dimensions or nominal dimensions. This is especially the case for older lumber because its larger dimensions can be a pain to integrate into a rebuilding project that also uses new lumber.

We suggest you measure the width and depth of a few pieces of the lumber in the building; and if your dimensions are within $\frac{1}{4}$ in. of the actual dimensions given in "Nominal and Actual Lumber Dimensions," on p. 240, you can refer to the lumber by its nominal dimension. If, however, the dimensions differ by more than $\frac{1}{4}$ in., refer to the lumber by its actual dimension. For example, if you measure a rafter and it is $1\frac{5}{8}$ in. by $5\frac{3}{8}$ in., it is within a $\frac{1}{4}$ in. of the actual $1\frac{1}{2}$ in. by $5\frac{1}{2}$ in. standard size, so call it a 2×6. If however, you measure the rafters as $1\frac{7}{8}$ in. by $5\frac{7}{8}$ in. (more than $\frac{1}{4}$ in. difference from the standard size), it's a good idea to tell your customers the actual size, especially if they aren't there to measure it themselves. That will let them know up front that they might need to resize this lumber to integrate into new construction.

It's also important to measure the length of lumber. Most softwood lumber today is sold in 2-ft. increments. So, it makes a big difference if your lumber length is 13 ft. 11 in. or 14 ft. 1 in. (the latter can be sold as a 14 footer, whereas the

*Always try to salvage the maximum length of lumber possible. The salvagers who worked on this building used a chainsaw to cut the 20-ft. 2×10s that spanned between the roof trusses; however, they located the cuts very randomly and, as a result, cut off 2 ft. to 3 ft. of usable lumber at each end. A real waste of high-quality lumber!*

former, depending on your market, might only be salable as a 12 footer—and you lose 2 ft. of value.). So, if you need to cut lumber, keep in mind these 2-ft. increments to minimize oddball lengths. The best practice is to measure lumber to the nearest 2 in. and round down the measurement so you never sell a piece shorter than advertised. If you trim the ends of the lumber (to remove nail-damaged or rotted ends), make sure you trim it square (use a radial-arm saw or circular saw with a framer's square). By the way, the exception here is that hardwood lumber is typically sold in 1-ft. increments.

## Assessing lumber condition

As with all materials found in a building, the condition of the lumber framing can vary. Obviously, you don't want to spend a lot of time and effort salvaging lumber that you discover is in poor condition and unusable. More important, because wood decay and insects can do serious damage to the structural frame of the building, your personal safety can be at risk if you can't recognize at least the obvious signs of problems. Over time, we have gained some practical knowledge about problems you will want to keep an eye out for as you look at the wood in your building.

Wood decay, termites, and carpenter ants are the main causes of wood degrading in a building. Wood decay fungi, which rot wood, require the presence of moisture. If the wood in a building is dry (below about 20 percent moisture content), the wood cannot rot. Moisture in buildings comes from one of two sources: external water (rain, snow) or internal water (plumbing leaks or accumulated moisture vapor from washing, cooking, and so on). Roof leaks are the most likely cause of unwanted moisture in a building, though plumbing leaks can be just as destructive.

*The timbers* in this barn show some water staining, a sign of a leak at some point in the structure's life. Closer examination by poking with an ice pick or knife blade will tell you if the wood is sound or not.

## Dry Rot: Can Wood Really Rot If It's Dry?

We've all seen the results of wood decay: a once strong, stout timber reduced to a pile of mush. Wood decay is usually caused by one of two types of fungi: white rot fungi or brown rot fungi. Both are equally destructive; and it's not so important to be able to identify the species, only its presence.

As living organisms, fungi need four basic conditions to survive: a food (the wood), moderate temperatures, moisture, and oxygen. Remove one of these conditions and wood will not rot. Wood treated with a preservative chemical such as chromated copper arsenate (CCA), creosote, or one of a host of others, eliminates the wood as a food source for the fungi (and termites). Also, wood fungi will not thrive in extreme low or high temperatures. However, most buildings are kept warm enough in the winter and summer to support the growth of fungi, so controlling temperature to stop decay is not practical. Removing the oxygen is also impractical. (It's interesting that wood can last centuries under water because oxygen levels there are low enough that decay fungi can't survive.)

Moisture is the one condition you can control to prevent decay. Wood must have well over 20 percent moisture content to support fungal growth, so wood used in construction (usually dried to 19 percent or less) will not decay unless moisture from some other source is present. The term *dry rot* is a misnomer and is technically incorrect; wood cannot rot if it's dry.

*It's important* not to overestimate yield when there are signs of decay in a building. The decay in this timber extended back over 2 ft., reducing the final yield.

*This Douglas fir flooring was attacked by termites but the more naturally durable redwood timber on which the flooring sits was not affected. The termites left it alone for an easier meal.*

Moisture is also required by two of the three most common types of termites—subterranean and dampwood. The third, the drywood termite, needs less moisture and can be harder to detect. All three are destructive, though the worst type of termite is the Formosan, a subterranean termite from China. Its range has spread from New Orleans, where it is believed to have entered the country, and can be found from Texas to Florida in the southern United States.

In addition to looking for decay and insect damage, it's always good to keep an eye out for other damage that may negatively affect your yield of lumber. Large splits, cracks, notches, or holes in the wood will usually mean the board is unusable or will have to be cut down in length or width to eliminate the damage. Some damage you can't avoid, some you can. When taking apart the lumber, try to minimize cracking and splitting by not prying the boards too forcefully or dropping them from height. This will

*The heart pine floor of this old textile mill got wet from a roof leak and buckled in many places, making salvage difficult and yields of reusable wood low.*

*The notches* in these timbers reduce the quality and value of the lumber.

*Under the dirty skin* of old beams can hide some very beautiful wood, as indicated by this remilled Douglas fir timber ready for installation in a new timber-frame house.

only reduce your yield and/or lower the value of the lumber you have worked so hard to harvest.

While we have emphasized the worst that can happen to the wood frame of a building, don't be discouraged by what at first looks like dirty old timbers. Very beautiful wood can lie under that grungy facade.

## Protecting your lumber

Once you've gone to all the trouble of removing the lumber from your building, you will want to denail it, remove any metal clips or hangers, trim any splintered ends, stack it in bundles, and store it properly. There are a couple of things to bear in mind here. First and foremost, take the time to remove the nails as close to the point of taking the lumber off the building as possible. Lumber with nails or other hardware attached to it is awkward and dangerous to handle, it won't stack evenly and the bundle will be unstable, the nails are harder to remove after they've bent over, and you will do more work because you will eventually have to unstack, denail, and stack again.

Second, resist the temptation to simply throw boards in a pile. Though it may seem expeditious while you're doing it, it will cost you extra time and effort in the long run, as you will have to untangle and restack the boards at a later time. In the meantime, the lumber is subject to damage.

*This is not the way to stack lumber. Take the time to remove the nails and stack the wood in an orderly pile.*

*If you generate* a lot of lumber, it's worth investing in a set of strapping tools. A set includes a tensioner (top left), which tightens the band, and a sealer (bottom left), which crimps the overlapping bands together.

*Properly stacking* and securing the lumber on the truck will ensure that your load gets to its new destination safely.

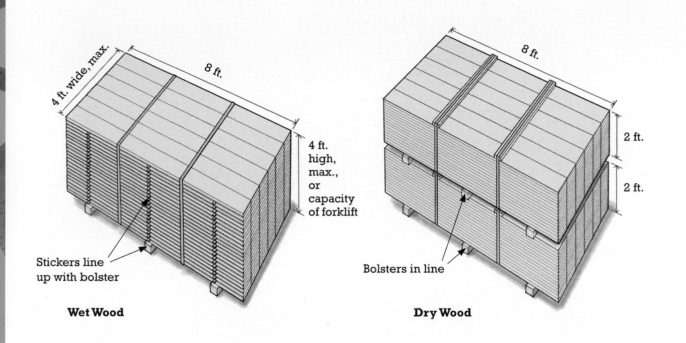

4 ft. wide, max.

8 ft.

4 ft. high, max., or capacity of forklift

Stickers line up with bolster

**Wet Wood**

8 ft.

2 ft.

2 ft.

Bolsters in line

**Dry Wood**

If you deal with a lot of lumber, it's a good idea to invest in a set of steel strapping tools to band your lumber. To hold together a large stack of lumber you need the strength of metal bands, not plastic. The last thing you want is a 2,000-lb. stack of lumber coming apart when it's over your head on a forklift or truck.

How you stack the lumber in each bundle depends on how dry it is. Most lumber coming out of a building is dry enough that you can lay each board one on top of another with no space in between. If the lumber is wet, it's best to "sticker" the boards so they can dry out. Stickers are small, uniformly sized sticks separating the layers of lumber, usually spaced about every 2 ft. for 1-in.- to 2-in.-thick lumber.

*The small sticks* separating the layers of lumber, called stickers, allow air movement for even drying if the lumber is wet. It's important when stacking lumber to make sure the stickers line up not only top to bottom, but also with the bolster (the larger 4×4 that supports the whole bundle and keeps it off the ground).

Finally, size your stack so it will fit on the forks of the forklift and the truck you intend to haul it on. Stacks about 42 in. wide are ideal. Don't layer too high, or your forklift may not be able to pick it up. A 2-ft.-high, 4-ft.-wide stack of lumber 8 ft. long can weight up to 2,000 lb.

*Don't make* your lumber bundle wider than the length of the forklift forks!

## Selling Your Stuff

If you aren't going to save the materials for your own reuse, you'll probably want to sell them. Placing a classified ad in your local paper under the building materials category is a good first step; and if you time it right, the ideal time to sell your materials is on site. If your site is on a busy road, putting up a sign (reading, for example, "Building Materials for Sale Here") can draw local customers. Also, some localities have Web sites for selling building materials. Your local government recycling coordinator should be able to help you find a locally run Web site.

It's best to establish a price before you advertise. Although some people are not afraid to make an initial offer, others are intimidated and will walk away if no price is given. Also, many people associate used building materials with cheap or inferior and expect rock-bottom prices. You want to be honest about what you have, but don't be afraid to sell your materials. You will attract more (and higher-paying) customers if the ad reads "For sale: Original unpainted golden oak trim from 1920s Arts and Crafts home" rather than "For sale: Old wood trim from torn-down house."

If the value of the material or item you are selling is high enough, consider consigning it to a local used-building materials resale center, antiques shop, or architectural salvage store. This provides storefront display, though it will cost you a portion of the sale price. Some deconstruction entrepreneurs have had great success with on-site auctions or sales. Typically, this is limited to materials that are easily soft-stripped. Items are marked with prices (or numbers,

if an auction), and the customer is required to remove the materials, using his or her own tools and help, by a specific date. These auctions require advertising and some thought on luring customers to the auction site. Also, you should check with the building owner or your insurance company to make sure existing insurance policies cover such an event.

Also, don't forget the power of the Internet. We have seen reclaimed building materials (as well as whole buildings for salvage) for sale on eBay®. Although shipping costs can prohibit the long-distance sale of many lower-cost building materials, Internet auction sites such as eBay or Craigslist℠ can attract local customers. Search using the keywords "used building materials."

If you are feeling charitable, another option is to donate the materials. Many organizations accept building materials. Habitat for Humanity ReStores, Goodwill Industries,℠ Salvation Army,℠ and St. Vincent de Paul℠ are some of the organizations that might be interested. Just contact them first to be sure they need or want what you have to offer.

*Kevin Brooks of Kevin Brooks Salvage in Philadelphia deconstructs unwanted buildings to acquire salvage both to sell and to use to build distinctive furniture.*

"When I first started salvaging, I would rescue things from the trash with no specific plan for reuse."

# Unbuilding the Inner City

## Kevin Brooks

As a child, Kevin Brooks wandered the abandoned brownstones and shuttered factory buildings along the Philadelphia waterfront. "I would try to imagine these buildings as they once were—elegant, functional, productive, and useful. I wondered about the people who lived and worked in these places so long ago and tried to relive their times."

As Kevin got older, he realized the potential of the material hidden under the litter and behind the weeds and imagined that someday somebody would "make the effort to preserve, restore, or re-create the beauty in them." As it happened, that somebody was Kevin Brooks.

Over a 10-year career as a senior policy maker in Philadelphia's Office of Housing and Community Development, Kevin led a double life. During the day he inspected old buildings, advocated for preservation, and oversaw renewal-style demolitions. At night, he rummaged through trash containers and abandoned buildings to retrieve discarded materials.

Though some people in his neighborhood affectionately referred to him as "Sanford," after TV's favorite junkman on *Sanford and Son*, it was clear that Kevin had a talent for seeing the treasure among the trash. For him, it had become a passion: giving new life to discarded building materials.

"When I first started salvaging, I would rescue things from the trash with no specific plan for reuse," Kevin says. "It was enough to surround myself with beautiful and interesting objects. And I enjoyed showing off to my friends the valuable things I got for nothing."

Kevin honed his craft and soon filled storage lockers, a garage, and a large delivery truck with his collection. At the point when his three-bedroom apartment was filled with old building materials, fixtures, and hardware, he decided his hobby had outgrown its part-time status and should become a full-time business. Kevin quit his day job and turned his collection into Kevin Brooks Salvage, a construction company based on the premise that if he could efficiently salvage the city's unwanted building materials, people would pay him for them.

Working out of an old 1865 brewery, Kevin began to build a business, develop his deconstruction expertise, accumulate inventory, and establish a following of customers drawn to his growing inventory of industrial artifacts and salvaged materials. Collaborating with interior designers, artists, and other customers, Kevin now integrates the salvaged materials into commercial use in local restaurants, retail shops, and professional offices. He also began a second business, Found Matter, a design company that makes furniture from salvaged materials.

"I have always derived so much joy from looking at old structures, discarded artifacts, and obsolete pieces of equipment and and posing the simple question of What if? The possibilities are endless," says Kevin. "I'd encourage anyone interested in deconstruction to try my little test, it's a thrill . . . . I find it incredibly rewarding to be part of an effort to help create economic opportunity by adding value to materials that are otherwise viewed as waste."

# GETTING STARTED

## Chapter 4

*It's a job that's never started that
takes the longest to finish.*

—J. R. R. Tolkien

Once you've determined that your candidate building passes muster and you've decided to take the plunge into unbuilding, it's time to think about getting started on the actual deconstruction process. Every situation is different, but the first step in any project should always be organizing the site. A few hours spent in planning and systematizing your activities can help you do the job efficiently, safely, and quickly. In this chapter, we talk about some of the site issues you'll want to think about before putting on your gloves and hardhat. We also give an overview of some of the tools you'll need.

## Organizing the Site

Developing a strategy for your particular site will minimize the disorganization that can slow work, create confusion, and make for an unsafe workplace. As a first step, it's a good idea to sketch a simple site plan that will force you to think about where various activities should take place, where equipment and tools will be stored, and where materials can be stockpiled. An important consideration in your planning should be, when possible, never to put something where it will have to be moved again. Nothing wastes time and money faster than moving piles of materials unnecessarily.

*Before work begins,* it's important to spend some time developing an efficient strategy for your unbuilding site.

## Establish paths for people and equipment

Begin by thinking about the most logical paths for the movement of people and equipment. Trucks will need to come and go to drop off and pick up roll-off trash containers and to load materials, so there should be an easily identified entrance and exit to your site. Your crew will also need to move about the site and throughout the building. As much as possible, try to keep paths for people and the movement of machinery separate.

*Removing some interior* non-load-bearing walls in this house established a clear path, allowing easy movement of materials from the back bedrooms to the front porch for loading.

*Allowing clear access in front of this house made it possible to load materials directly off the porch using a ramp.*

*Having a parking area that's close to the job site is convenient for workers, but it can be a problem if cars need to be moved regularly to make way for equipment or material transfer.*

Try to establish a direct route from the place the materials are generated to the denailing, loading, and storage areas. You need to allow some flexibility, however, as you may want to have multiple exits out of the building for movement of materials. There are many ways to get materials out, from carrying them out the door to lowering them out a window to cutting a window opening down to the floor and creating a ramp. So don't feel your options are limited—with a little planning, you can make multiple routes for people and loads and do it efficiently.

Establish a parking area for your crew and visitors that doesn't get in the way of the movement of equipment. It's a time waster if your crew has to stop work throughout the day to move cars to accommodate equipment. When possible, keep the cars far enough from the materials processing area to prevent tire punctures and scratches (or worse) caused by moving lumber around.

As you're thinking about the movement of equipment, consider and plan for the weight of trash containers and the trucks that deliver them. Determine if (and where) a septic tank

*It's important to locate the job site trash container to allow easy access without hindering movement of workers and equipment.*

might be located on the property and flag its location prominently. Make sure you have sufficient overhead clearance for drop-off and removal (in most cases, the truck will need to back up to deliver and pick up) and room to open and close the door of the container (allowing for a wheelbarrow ramp, too). You want to avoid a delivery truck stuck in the mud or, worse yet, a downed power line. Besides being dangerous, both can be expensive and time-consuming.

*A long, open drive with plenty of head-room is required for the unloading and loading of a roll-off container.*

*Allow enough room in front of the container for the doors to open and to accommodate a wheelbarrow ramp (not shown).*

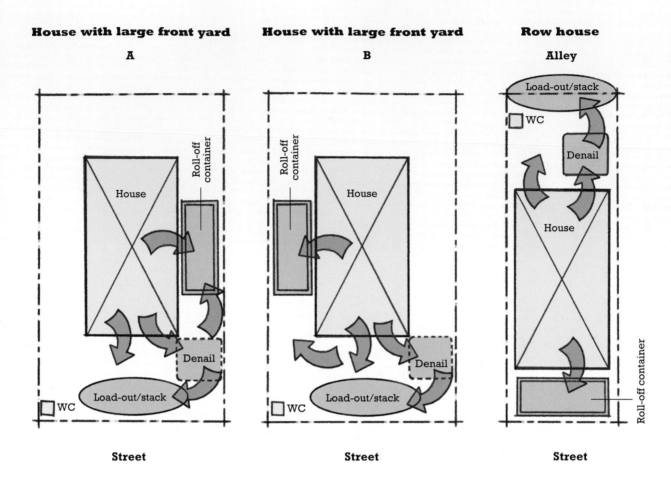

**House with large front yard**

**A**

**House with large front yard**

**B**

**Row house**

**Alley**

Street   Street   Street

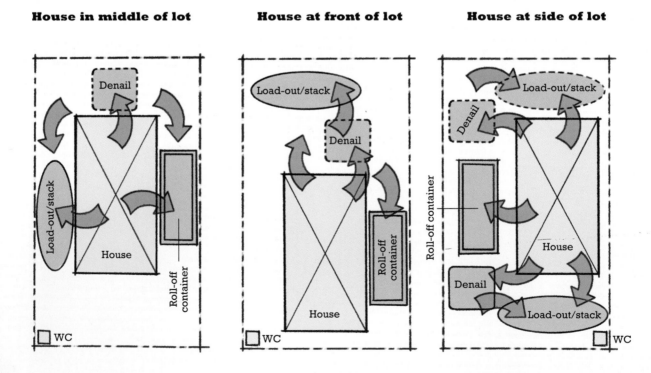

**House in middle of lot**

**House at front of lot**

**House at side of lot**

## Minimize the movement of materials

As you look at the site, try to minimize the distance materials have to be moved by hand. The sooner you can get materials onto a truck, trailer, pallet, or forklift for transfer the better. This is especially true for higher-value items and those that need special handling, like fixtures and windows. These will often be the first items removed and the most valuable, so you want to get them off site as soon as possible to minimize the risk of theft, breakage, and weather damage.

If you're working on a very tight site where there's only one area for material processing, loading, and a roll-off container and everything needs to flow along one path and through one doorway or other opening, you'll want to start your deconstruction as close to that doorway as possible and work away from that point toward the inside of the building. By removing things and cleaning up as you go, you are clearing out and opening up the space, easing the flow of people and materials from the starting point into the core of the building. In simple terms, you are trying not to paint yourself into a corner. Obviously, for second floors, you'll want to leave access like stairways in place and floors intact until everything else is removed and you're ready to move to the lower level.

## Materials processing and storage areas

*Materials processing* is just a fancy name for what you do to the salvaged items once they are removed from the building. It might be taking out

**HERE'S A TIP**

### The Best Spot for a Roll-Off Trash Container

It's always best to locate your trash container as close to the building as possible. This will allow you to push roofing right off the building into the container with no added labor. Also, if you situate the container under or alongside a window, you can move debris directly out, once the window has been removed. Note that if you're working more than 20 ft. above grade, U.S. Occupational Safety and Health Administration (OSHA) regulations require enclosed chutes to funnel debris down to the ground or into a container.

*It helps to have clear, open pathways for the safe movement of materials. Here, windows were taken out and a wide opening created in the front of the house so materials could come out onto the porch and directly into a waiting truck.*

**HERE'S A TIP**

### Make a Plan

Organize your site to minimize the movement of materials. You want to handle the materials as few times as possible. Try to minimize the movement between the following steps:

1. Removing materials
2. Processing (denailing, trimming, and so on)
3. Loading
4. Unloading

nails, trimming, taping parts together, shrinkwrapping, bagging, or boxing. An organized materials-processing area is essential for keeping materials sorted properly and ensuring their safe movement. Try to envision the various items that will come off the building—such as 2×4 wall studs; 2×6, 2×8, and 2×10 rafters and joists; 1×s, and plywood or other sheathing—and how big a stack they might make after they've been denailed and trimmed.

Lay out areas to store these materials with enough room around them to safely maneuver, and be prepared to tie down tarps and provide truck or forklift access. Remember to locate stacks of lumber so a forklift has enough room to get its forks under the stacks and move them around. When locating the processing area, try to minimize the distance you have to carry materials that contain nails.

*Even a small house* can generate a lot of materials that must be segregated, stacked, and inventoried.

*Pile recyclables for pickup* at the end of the project in a location that won't get in the way of the movement of people or materials.

Decide on areas for piles of recyclables, which should be off the beaten path but accessible for loading at the end of the project. Because some materials, such as metals, will come off the building in a variety of shapes and sizes (conduit, gutters, metal lath, wire, cast-iron plumbing fixtures, roofing, flashing, aluminum windows and siding, and so on), a pile on the ground may be as organized as you can get.

## Denailing station

You should always plan for a denailing station if you are salvaging siding, framing lumber, or other items that will come out of the building with nails in them. A denailing station can be as simple as two sawhorses to hold the lumber off the ground and a plastic tarp underneath to catch pulled nails. From a materials handling standpoint, its best to keep denailing close to the building for two reasons: (a) to minimize the handling of lumber with nails in it, and (b) to keep loose nails away from the path of truck and car tires. Of course, you don't want to be so close to the building that the crew at the denailing station is in danger of being hit with falling objects or is in the way of deconstruction activities. If you anticipate that you'll be trimming a lot of lumber, you may want to locate the processing station near the trash container for easy disposal.

Establish as direct a path as possible for materials headed to either the trash container or the loading area. If materials need to be processed (denailing or trimming), locate an area for this between the building and the loading area.

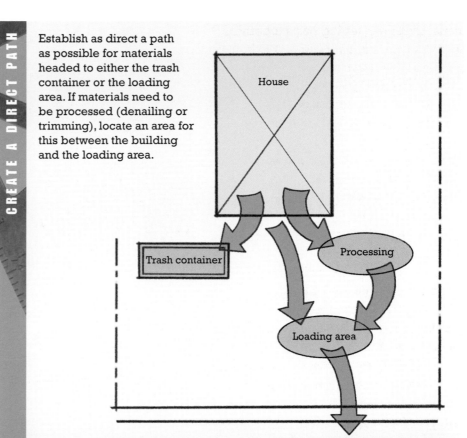

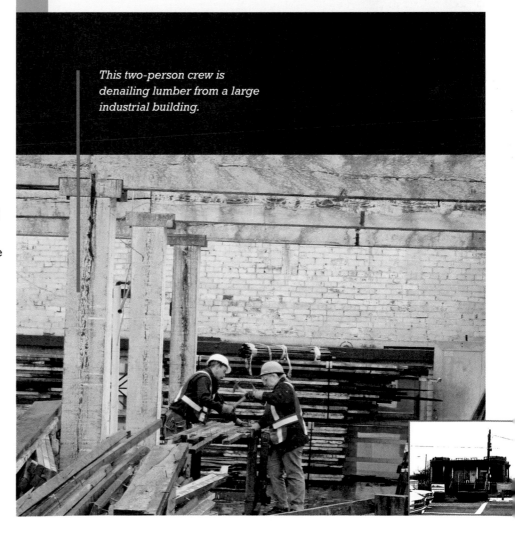

*This two-person crew is denailing lumber from a large industrial building.*

*Locate the denailing* station in the shade, and your crew will be eternally thankful on hot and sunny days.

## Site safety and security

As on any construction site, you need to think not only about the safety of your crew but also about the safety of neighbors, passersby, and anyone who might wander onto the site. If the site is not already fenced in, it's a good idea to establish an outer boundary of safety by erecting rented chain-link fence (expensive), bright orange snow fence (cheaper), brightly colored caution tape (cheapest), or other barricades where gaps exist. Posting a job-site sign at the main entrance that informs the public that a deconstruction project is taking place and posting a *copy* of the demolition permit helps outsiders understand that caution is warranted. Also, posting a sign that reads "Danger: Hard Hat Area" is a good warning to outsiders and a reminder to crew members to wear hard hats.

Yellow "caution" tape or red "danger" tape should be used to warn of an imminent hazard on site: for example, to cordon off an area to prevent someone from walking where roof materials are being dropped to the ground (red tape), to indicate where heavy equipment is in use (yellow tape), or to block off the bottom of a stairway if dangerous activity is going on upstairs (red tape). Be sure your crew knows the meaning of each color.

*The entrance to this unbuilding site is highly visible, with a job-site sign and an orange snow fence that was pulled across the driveway at the end of every workday.*

*Red "danger" tape was used to cordon off the front entrance of this house during roofing removal. All crew members should know not to cross into a hazardous area unless they have a clear signal to enter from the workers creating the hazard.*

## Dealing with vegetation and wildlife

If your site is otherwise suitable, but somewhat overgrown, you might need to remove trees and plantings to gain better access and working space (get permission from the owner first). If there are a lot of weeds around the building, it's a good idea to get rid of them too. Although this takes time away from salvaging, it helps provide safer footing, which is especially important for somebody walking with an armful of lumber.

If there are heritage, or otherwise significant, trees on the site, you'll need to consider how to protect them. Trash containers and heavy equipment can damage tree roots, so, where possible, try to keep beyond the drip line of the tree. The drip line is defined as a circle around the tree whose radius is the distance from the trunk to the outermost branch. You may need to remove or trim some trees to ensure the safety of workers— for example, when branches overhang a roof.

*If there are bushes growing close to the house, you'll need to take them down to allow access to the siding for removal.*

It's also a good idea to keep an eye out for nesting birds or other animals that might be living on site. If the house has been abandoned for some time, you're likely to find less than cuddly critters, such as raccoons, snakes, squirrels, or bees, protecting their turf. You should also be aware of any native plants that are protected in your state or county and take care not to injure them.

*A nest of California quail eggs was found just in front of this house to be deconstructed. We put up a warning tape around the nest and carried on with the project. Because the crew was careful, the eggs hatched and the quail family moved on no worse for wear.*

## Creature comforts

From a health and safety standpoint, drinking water is an essential item to have well stocked at your job site, especially in hot weather. A nearby tap is perfect (assuming the water is potable), but otherwise bring in cases of bottled water or a refillable 5-gal. plastic water cooler. If you are handling lead-based paint (LBP), we recommend using disposable paper cups for drinking water to prevent the spread of lead dust to water bottles or cups you may take back home.

It's also important to have a washing station or a larger barrel and soap for hand washing before lunch and after work, especially if you are dealing with materials coated in LBP. If getting washing water to the site is a problem, at the very least have "wet wipes" available for cleanup. A whole-house deconstruction that takes more than a few days may require a portable toilet. If so, locate it downwind and not too close to the break/eating area.

If you will use a generator for power, you might want to locate it in a spot that is away from the main work area, because of the noise. Wherever you put a roll-off container, a denailing station, or even a portable toilet, be conscious of neighbors and the effect the noise, activity, increased truck traffic, and smells will have on their lives.

A comfortable rest area can make an otherwise difficult job easier. If possible, set up chairs and a table to make lunch a more enjoyable experience. We have seen temporary tents and awnings erected at some deconstruction sites, and they serve well to provide shade and a communal meeting place.

*A hand-washing station* and a portable toilet are necessary items on the unbuilding site. The water in a hand washing station is not for drinking.

## On-site communications

If you have a sizable crew, it's a good idea to have more than one cell phone on site and a posted list of key telephone numbers. Good communication is an ever-present issue. On a large project, you are constantly communicating with others: outside laborers, subcontractors, potential buyers of materials, waste and recycling haulers, property owners, and so on. Depending on the size of the project, plan on spending a good amount of time on the phone if you want things to go smoothly. Call people early in the process and often so that your scheduling and efforts are not impeded.

It's often hard to get other workers' attention on a noisy job site, especially if they are concentrating on what they are doing. It's always prudent to wait for a worker to pause, especially if he or she is doing something potentially hazardous, like swinging a sledgehammer or running a circular saw. Also, never enter a hazardous area without clearance from someone working in that area. You wouldn't want to walk under a big timber beam whose support is being cut out from under it.

*A lunch tent* was set up for the duration of this military building deconstruction project.

### Turn Off the Power

Although it may seem convenient to leave electrical power on and selectively shut off circuit breakers at the house subpanel or main electrical panel, we strongly discourage it because you should never assume that circuits are wired as labeled. Also, novice electricians may have added circuits in places not expected during earlier remodels. In any case, if you leave electricity on in the building, you have to assume that each outlet or fixture is hot, increasing your risk and anxiety. In the end, it's much safer to disconnect the power to the building at the street and use a generator or to get a temporary power panel installed on a job-site pole (just as on a construction site).

*It's much safer to install a temporary power pole and panel or to use a generator for job-site power rather than using the building's electrical system.*

If you have a large crew and you need to get everyone's attention, consider buying an air horn. They are cheap, loud, and often refillable for reuse. Establish a common meeting point (perhaps the rest area) and a simple signaling system to get everyone's attention (for example, one blast means it's time for lunch break or a meeting; two blasts means there's an emergency, so stop working and proceed to the meeting point). Use them judiciously, to avoid complaints from neighbors.

## Disconnecting utilities

For a whole-building deconstruction, the best strategy is to have all utilities disconnected before any work starts. Most municipalities require a demolition permit, and most demolition permits require utilities to be disconnected. This process is normally handled by the local gas or electric utility, telephone or cable company, or municipal utility, who will disconnect the service at the pole (electric, cable, telephone) and the street (gas, water).

Usually, the electric utility will disconnect existing electrical service at the distribution pole to protect the wires from damage and people from electrocution caused by equipment that might hit the wires. Disconnecting at the pole is best because you won't have to worry about overhead lines, and you'll have more room to bring in trucks and other equipment with high-clearance requirements.

Natural or propane gas disconnects are crucial for obvious safety reasons. Selective disconnection of gas lines is unnecessary because you will have no use for gas during deconstruction. Have it shut off at the street. Telephone and cable TV wires pose little risk, but should be disconnected at the pole by the appropriate utility.

Shut off water as close to the source as possible. It's easier to run a hose from a temporary hookup at the street or from a neighboring building than to worry about finding shutoff valves in the house to selectively disconnect. Having a connected water source is not essential because a cooler and washing station can take care of drinking and hand-washing needs, though a water source is handy if you want to wet down dusty materials, such as blown-in insulation.

Septic tanks are common in some communities, especially in rural areas, and typically require a close-out permit and certification from the local government building department or public health department as part of the demolition permit. You should figure out if the house is on a septic tank before the job begins and plan for the cost and scheduling to have it pumped out and capped or removed.

## Tools for Unbuilding

Although, as quoted below, WD-40® and duct tape come in handy on any deconstruction site, they aren't the only tools you'll need. Many of the same tools you find on a construction site come in handy when unbuilding, but because you are taking the structure apart, the emphasis is on tools that pry, pull, cut, and pound. Unlike a carpenter, you won't have to do much in the way of detailed measuring (except possibly at the lumber trim station), so squares, rules, scribes, and levels aren't commonly used tools. You could spend a small fortune on tools, but you don't need every tool in the hardware store to get started and you can begin a soft-stripping project with a modest investment. The best advice we can give is to use the correct tool for the job at hand.

We've put together a list of tools that we consider essential for deconstructing a house (see "What You Need" on p. 120). Obviously, there's no sense in buying every tool on the list if you plan on doing this only once. We've also found through experience that it makes more sense to purchase fewer high-quality tools than many low-quality tools. A higher-quality tool lasts longer, is typically more comfortable to operate and hold, and thus is safer. That said, you don't have to go overboard and spend $200 for a titanium framing hammer when a $25 model is just fine for pulling nails.

*Many of the tools you'll use on an unbuilding site are those that pull or pry something apart. A hard hat is also strongly recommended.*

*One only needs two tools in life: WD-40 to make things go, and duct tape to make them stop.*
**—G. Weilacher**

## What You Need

In our experience, you need a relatively small number of tools if you're planning on doing soft-stripping and removing little or no structural materials. On the other hand, if you are contemplating a whole-house deconstruction, you'll want to broaden your tool arsenal. All the tools you'll need for soft-stripping are usable in a full deconstruction so here is a basic list of tools necessary for unbuilding (organized alphabetically, not by preference). See p. 242 for additional tools, which you can add as necessary.

### Basic Tool List

- Box cutter (or utility knife)
- Cat's paw (or nail puller)
- Chainsaw, small, 14-in. bar (only if cutting lumber in place)
- Circular saw (only if cutting lumber)

- Cordless drill or screwdriver, 7v or larger
- Extension cord(s)
- Hammer, framing
- Ladders
  - Extension (only if accessing second-story windows or other elevated items)
  - 4 ft. (comes in handy)
  - 6 ft. (a good starter size)
- Nippers (if removing reusable trim)
- Pipe wrenches (if removing sinks or other plumbing fixtures)
- Pliers
  - Channellock® (if removing plumbing fixtures)
  - Regular
  - Vise-Grip®
  - Wire-cutting, insulated
- Prybar (flat blade)
- Reciprocating saw (such as a Sawzall®)
- Sawblades, wood- and metal-cutting

- Sawhorses (several pairs, as needed)
- Screwdriver bits, #2 size, 2 in. or longer, Phillips® and slotted
- Sledgehammer, 12 lb.
- Tape measure, 16 ft. or 25 ft.
- Tool belt
- Wrecking bar, 24 in. (longer if bigger timbers)

### Other Useful Items

- Brooms and rakes
- Buckets, 5 gal.
- Garbage bags, heavy contractor bags
- Plastic bags, zip-closing
- Shovels, flat bladed
- Wheelbarrow

### Tools for Safety

In this chapter we are focusing on the tools used for physically taking apart the building. In chapter 5 we focus on safety, so don't forget that there are safety items, described there, that need to be part of your tool collection.

*You'll find countless uses for a cordless drill or screwdriver—removing everything from light fixtures and kitchen cabinets to hinges and switch plates— and you'll save the bother of dragging extension cords around the job site.*

**Buying Cordless Tools**

Most manufacturers offer combination kits with several cordless tools (for example, drill, circular saw, and reciprocating saw). These kits typically include only two batteries and a single charger. Although they are great for one person, you'll quickly end up with two dead batteries and at least 15 minutes to wait to recharge if more than one person uses tools from the kit at the same time, so it's a good idea to buy extra batteries and chargers. We also suggest that you buy different types of rechargeable tools from the same manufacturer (of the same voltage and charger type!) so that you will have plenty of interchangeable batteries and chargers.

## Corded or cordless?

In addition to hand tools, a range of corded and cordless power tools are available from a variety of manufacturers. In the last several years, both the variety and the power of cordless tools have increased dramatically. Powerful 36v cordless tools are now available, up from 9v only a few years ago. Also, battery life is getting longer, while charging times have shrunk dramatically.

Whether to use corded or cordless is a matter of personal choice, though we recommend you consider a heavy-duty cordless screwdriver or drill for removing cabinets, fixtures, and anything else that's screwed down. Cordless obviously avoids the need to drag extension cords around the job site, though even with cordless you cannot operate without power because of the need to charge the batteries regularly. (Always buy a cordless tool with an extra battery so one is charging while you are using the other.) Cordless tools are ideal when going up and down ladders or working in crawl spaces or other tight environments. They also work well on platforms, overhead lifts, and other awkward places where space is at a premium. And when you constantly move a tool from place to place, a cord would be a trip hazard. It's hard to survive without a cordless drill or screwdriver, though we prefer a corded circular saw, especially when we're doing a lot of lumber trimming.

*A corded circular saw is the best choice at a heavily used lumber trimming station, though a cordless saw can be handy when working at height or in other situations when the cord would get in the way.*

## Pulling and prying tools

A hammer is probably the simplest and most useful tool on the deconstruction site. A standard hammer is typically manufactured from wood, tubular steel, fiberglass, or solid steel. Because of all the prying and pulling required, we prefer the solid-steel shank of a framing hammer (Estwing® is a good choice). Its straight claw is better than a curved claw for getting into tight locations.

If you need a heavy hitter, a sledgehammer is the way to go. Sledgehammers work well for knocking apart nailed sections of framing, such as a wall stud from the bottom plate, and have myriad other uses. Weighing from 2 lb. to 20 lb. they can wear you out quickly, so select a size appropriate for the job and the size of the person using it. A 12-lb. sledge is a good all-purpose weight to start with.

In addition to a hammer, you'll need other tools for pulling nails or prying pieces of lumber apart. The basic ones are the prybar and wrecking bar. A flat-blade prybar is an essential tool you should always keep in your toolbelt. The Wonderbar® (from Stanley®) is a commonly found brand (the Superbar® from Vaughan, Handybar from Estwing, and some generics are similar). These bars are relatively light but stiff enough that you can beat on them with a hammer. Unlike a larger wrecking bar, they are flat in profile and do minimal damage when prying off trim and other more delicate items. They also work well for pulling nails in tight places.

A wrecking bar is also an essential starter tool. We prefer some of the newer-profile wrecking bars with a 90-degree angle head to the traditional crowbar, because the crowbar's tight bend forces you to pull through a much longer arc to remove a nail. (This is the same reason we prefer a framing hammer over a claw hammer.) Wrecking bars vary in length from about 12 in. to 48 in. and are made in a variety of shapes to develop leverage for different situations. A good starter size is 24 in., which is light enough to use to pull nails overhead or from a wall. For prying bigger pieces of lumber apart or removing larger nails, you'll need a bigger wrecking bar.

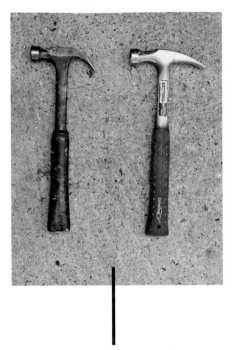

*Hammers come in a variety of lengths, weights, claw types, and handle types. Because of the abuse taken on a deconstruction site, we prefer a hammer with a solid-steel head and shank. The straighter claw of a framing hammer, at right, is a better choice than the curved claw of the claw hammer, at left.*

*A 12-lb. sledgehammer is a good size to have in your tool collection.*

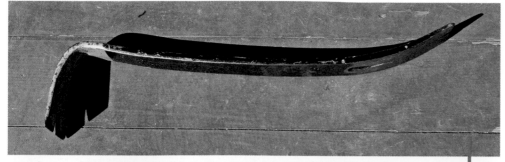

*The flat-bladed prybar,* available from many manufacturers, is an essential tool for deconstruction. Pair it up with a hammer, and you'll have a combination you'll use more than any other.

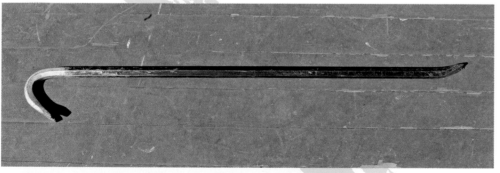

*A crowbar is a useful tool,* but we prefer the flatter profile of a wrecking bar (shown below). Several manufacturers make wrecking bars, and they are widely available. Check your local home improvement store.

## Grab the Head

For situations in which you can access only the head of the nail, a cat's paw fits the bill. Use your hammer to pound it underneath the nail head. Obviously, you wouldn't want to use this tool on nice trim because you'll damage the wood when getting a grip on the nail.

*A cat's paw is designed to be pounded under the head of the nail.*

*The cat's paw is a useful tool for pulling nails at a processing station. Placing a piece of scrap wood under the bar provides extra leverage in prying the nail.*

## Cutting tools

A reciprocating saw is another essential tool; one commonly known brand is the Sawzall from Milwaukee®, but this type of saw is made by many other manufacturers as well. The blade of the saw fits into tight spaces and cuts nails holding in windows, doors, and built-in cabinets. With the appropriate blade, you can cut metal or wood. Blades rated for "wood with embedded nails" are handy if you are cutting though the occasional nail, because you won't have to change back and forth from a metal blade to a wood blade. Although a basic model works fine, higher-end models have some nice features, such as variable speed, a longer stroke, orbital action, and swivel handles.

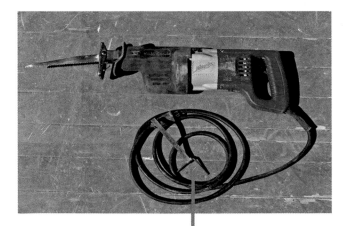

*A reciprocating saw is an essential tool on the deconstruction site for cutting nails and wood, especially in tight places.*

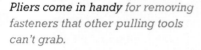
*Pliers come in handy* for removing fasteners that other pulling tools can't grab.

*Nippers and Vise-Grip pliers work well to pull finish nails from trim.*

While a reciprocating saw works fine to rough-cut lumber, a circular saw is the preferred tool if you are trimming lumber to length for stacking. Also, don't overlook the usefulness of a small chainsaw for cutting lumber, especially if you find yourself taking down a barn or other building with heavy timbers. They are less forgiving of nails than a reciprocating saw but cut much faster and don't bind as easily on large lumber.

## Miscellaneous tools

There are a number of other tools you'll need to round out a basic tool set. A tape measure is always useful; we like to keep a 25-ft. length in our tool belt, but it's also nice to have a 100-ft. tape if you want to make longer measurements, say, of the exterior building dimensions for estimating material quantities. You'll need various types of pliers, including a regular pair for gripping all manner of items, insulated wire-cutting linesman pliers for cutting away the electrical wires you will inevitably run across, Channellock pliers for unloosening waste traps under sinks, and Vise-Grip pliers for pulling out broken-off nails and other hardware that the prybar can't grab. It's also nice to have a pair of pipe wrenches for removing iron gas line from furnaces, water heaters, and other plumbing fixtures.

High-intensity work lights can be helpful in rooms that don't have enough natural light. Also count on buying a couple of contractor-grade extension cords. A utility knife (box cutter) is a good tool for all around cutting. Don't forget a tool belt to carry your hand tools (and develop the habit of putting all your tools back in it).

## Pulling Nails

Most interior trim is nailed with finish nails. A good method for minimizing damage to the front face of the trim is to pull the nails out from the backside. A pair of end-cutting nippers (one type is made by Channellock) works well, though you might want to dull the cutting edge some with a file so the nippers don't cut the nail before it can be pulled.

## Sawhorses, wheelbarrows, and cleanup tools

Sawhorses are invaluable on the job site. Inside a building, they can be used for stacking wood, trim, sheathing, paneling, and flooring. Using sawhorses minimizes bending, both for the person removing the material as well as for the person taking the materials for processing or storage. Place the sawhorses against a wall or in the middle of the room, but keep them away from entries and pathways.

Sawhorses are essential at the processing station so that lumber can be denailed and trimmed at a comfortable height. You can either buy sawhorses or build them yourself from 2×s. We like the painted metal variety: They are foldable for easy transport and will support a significant amount of weight, though you might want to screw a length of 2×4 on top of each one so you don't saw into metal with your trim saw.

A wheelbarrow can save you a lot of backache when catching or moving debris, especially dense drywall or plaster. By placing the wheelbarrow underneath the drywall removal area you can catch the drywall as it comes off the wall so you don't have to bend over to clean it up. Make sure you buy a heavy-duty model.

*We like these metal sawhorses because they are collapsible and will suport several hundred pounds of materials. If you want to stack particularly large piles of lumber, you'll want to purchase heavier-duty sawhorses.*

*Gravity is your friend. Place a wheelbarrow against the wall so you can drop lath and plaster directly into it and avoid having to stoop over to pick it up off the floor.*

A flat-bladed shovel is good for cleanup and works well for removing lath and plaster (more on this in chapter 6). You might also want a round-bladed shovel for other digging needs. Although they may be hard to come by in warm climates, snow shovels work well for pushing drywall and miscellaneous interior debris. A larger scoop shovel makes quick work of moving debris piles, if you can handle the weight. A heavy-duty garden rake is good for raking up debris, and you'll need at least one heavy-duty push broom for cleanup.

## Ladders and scaffolding

Unless you are 7 ft. tall, you'll probably need a stepladder to reach fixtures, windows, and other high-up items inside the house. A quality stepladder is a good investment; we recommend one that's 6 ft. to start, though a 4-ft. model also works if you mainly do inside work. You'll need an extension ladder for removing items such as window trim on the exterior of the house. Ladders are sold in three grades, which represent intended use and weight limits. Buy the best ladder you can afford. If you have a lot of trim or other material to remove up high, it may be worth renting or purchasing scaffolding.

*A flat-bladed shovel not only is useful for shoveling up debris but also makes a great tool for removing plaster.*

*Ladders are essential for even the smallest unbuilding jobs, and scaffolding can be a labor saver on most projects.*

## Specialty tools

In addition to the basics, there are a number of specialty tools to consider adding to your list. Although standard wrecking bars work to pry boards off the building and pull nails, other tools have been developed to make these jobs easier. Such tools work well in specific situations, and every deconstructor we know has a personal favorite. The two-clawed wrecking bar (made by Fulton), the wrecker's adze (from Vulcan®), the Grizzly® Bar (from Klein Tools®), the Board Lifter made by Metcalfe Roush, and the Extractor™ nail-pulling pliers made by Jefferson Tools are all useful tools. (See Resources on p. 243, for manufacturers' addresses.) We suggest you buy the basics first and then add the more specialized tools as budget and project type dictate.

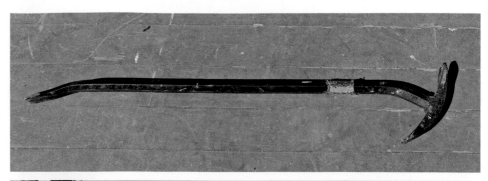

*The slight bend* built into this double-clawed wrecking bar allows you to pull nails from different angles.

*The Extractor* pulling pliers have a unique design that increases the gripping force as the tool is rotated.

*A wrecker's adze,* manufactured by Vulcan Tool Co., is useful for removing flooring.

*A Grizzly Bar,* manufactured by Klein Tools, has a long handle and wide blade for heavy-duty prying.

*The Board Lifter (far left) and the Flooring Lifter (left), both manufactured by Metcalfe Roush Forge and Design, are specifically designed to lift subfloors and finished floors, respectively. The angle of the Board Lifter's handle is adjustable.*

*The Nail Kicker, marketed by ReConnx, is a pneumatic nail remover that's invaluable on an unbuilding site. The barrel of the gun slips over the point of the nail, which is shot out when the trigger is pulled.*

One of the best innovations to reach the deconstruction industry is the Nail Kicker®, a handheld pneumatic nail remover. Although it requires an air compressor to use, this tool greatly speeds up production at a denailing station. To use the denailer, you place the metal tube on the business end of the gun over the point of the nail, pull the trigger, and the nail shoots out of the board. The nail comes out with a great amount of force and will ricochet off hard surfaces, so it's a good idea to lay down some carpet or shoot the nails into a padded garbage can.

*A rolling magnetic sweeper* is helpful for cleaning up stray nails that accumulate on the job site, especially around the denailing station.

## Tool storage

Tools left at a job site have a habit of disappearing. To protect your investment, use a job trailer, secured job-site toolbox, or lock box on your truck. It's usually not a good idea to leave anything on site, though in a pinch a lockable on-site roll-off container can be used for tool storage. Another option is to store tools in an adjacent lockable building, if one is available. To avoid losing tools, it's a good idea to check the job site at the end of each day because it can get pretty messy and tools are easily lost in the debris and piles of materials. If you have a crew, a checkout list for tools or a trailer with labeled places for each tool can quickly help determine if anything is missing at the end of the day.

## Tool rentals

Most tools you'll want to buy, but some are either too expensive or are used too infrequently to make it worthwhile. In that case, the local rental yard can help. Before your job starts, find out if the rental yard has the items you need on the dates you need them (and also find out if you have to pick them up or if they deliver).

A skid-steer loader can come in handy on big jobs and is relatively easy to run with some training. Many rental places will allow you to test drive and get used to the controls on their site before you rent. Manlifts,

*A job-site toolbox* is a handy and secure way to store smaller tools.

*A skid-steer loader* can be a real labor saver for moving materials.

*A long-reach forklift is a handy machine for moving materials from higher locations and for loading and unloading. This fork-lift is loading large timbers from an old warehouse onto a truck.*

long-reach telescopic forklifts, and scissor lifts also come in handy, especially for large timber-salvage projects. Because of costs, try to concentrate any work that requires a rental machine into as few days as possible. Some rental yards charge the same rate for a Friday to Sunday rental as for a single-day rental, so if you have the option of working over the weekend you may have more flexibility with the machine. Be aware that most machines have a run clock, so you may be charged extra for run time over the single day limit (typically 8 hours).

One last thing to think about is the vehicle you are going to use to haul yourself, your tools, and the materials

*A scissor lift can come in handy on big jobs, as long as you have a smooth, stable surface to drive on. Using a lift made it much easier to remove the lumber purlins from this warehouse.*

you harvest. It is impossible to do this kind of work without a pick-up, van, or trailer. If you are serious about moving a lot of materials, think about getting a 1-ton truck and a trailer. A fifth-wheel trailer is the best for big loads. If you have to rent a van, you can try to set up a lockable roll-off or some other way to secure your tools at the site, and reserve the use of the rental to drop everything off at the beginning and pick it all up at the end of the project.

*Linda Lee Mellish and Lara Kelly sell
salvaged architectural materials
and other reusable building materials
at ReStore in Philadelphia.*

"Doors are one of our most popular items.
We have had customers plan
an entire renovation around one
beautiful door."

# Architectural Salvage for All

**Linda Lee Mellish**
**Lara Kelly**

During its first year of operation, ReStore, the brainchild of Linda Lee Mellish and Lara Kelly, saved 30 tons of usable building materials, fixtures, and architectural elements from the landfill. ReStore of Philadelphia (unaffiliated with Habitat for Humanity ReStores) is unique in that it doesn't purchase any inventory; instead, deconstruction is done in exchange for the unwanted items. This not only saves the contractor and home-owner money but also makes the usually pricey architectural salvage affordable to everyone.

Visitors to ReStore, a three-story brick knitting mill built in the 1870s, find a little of everything— intricate faucet fixtures, pastel toilets, beveled-glass windows, and massive paneled doors. "Doors are one of our most popular items," says Linda Lee. "We have had customers plan an entire renovation around one beautiful door."

The paths of these intrepid women crossed when Lara bought the shell of a house in Philadelphia. Wanting to use recycled materials as the basis for the project, Lara enlisted Linda Lee to help her integrate used objects with new ideas. During her 20 years as a carpenter, Linda Lee had become known not only for her building expertise but also for her commitment to environmental stewardship. Lara was an art department coordinator on several Hollywood feature films, which often required period-correct props. Rather than have to make them from scratch and then distress them, she searched for authentic materials but was frustrated with the lack of sources in Philadelphia.

"The only salvage available was of very high-end antique materials," says Lara. This lack of affordable reused building elements was doubly frustrating given the age and history of Philadelphia and its diverse population. Rebuilding the interior of Lara's house was such a success that the two decided to extend their collaboration into a business.

"We have a unique business model," Lara says. "There are salvage companies all over the country but they focus either on low-end or very high-end users. We started this business to make architectural salvage available to everyone while helping reduce the flow of usable materials to area landfills." ReStore's business plan was born of necessity, according to Linda Lee. "We had lots of sweat equity to invest, but no capital."

After 3 years of operation, hard work remains the key investment. Linda Lee and Lara still do the dirty work, closing ReStore on Mondays and Tuesdays to dismantle and stock their salvaged goods. Of all the items passing through ReStore's doors, Linda Lee's personal favorite is a set of servant's bells salvaged from a house in Germantown, Pennsylvania. The different tones of the bells told servants which floor of the three-story dwelling required assistance.

The servant's bells count for only a tiny fraction of the 30 tons salvaged that first year. Linda Lee and Lara, however, measure their success not just in weight and money but also in terms of what is needed to keep up with demand: a larger building, a loading dock, and a dismantling crew.

# SAFETY AND ENVIRONMENTAL HEALTH

# Chapter 5

*The way I see it, if you need both of your hands for whatever it is you are doing, then your brain should probably be in on it too.*

—**Ellen DeGeneres**

**N**ot unlike building a house, taking one apart can be hazardous, and construction is consistently ranked as one of the most dangerous professions in the United States. Injuries that occur on a job site can be prevented, but it requires each person to take safety seriously and to use his or her head. To keep yourself and others around you safe, it's important to constantly think about the task you are doing and whether you are performing it in the safest manner. Safety amid the ever-changing conditions of a deconstruction site also means fostering a collective awareness so that everyone on the project is not only watching out for himself or herself but is watching out for everyone else as well.

Of course, not everyone can be expected to be safety savvy right out of the gate, so somehow, from some source, everyone who is on the site needs to learn what safe practices are and to be regularly reminded of them. We suggest discussing safety (and environmental health) at the start of the project, with a follow-up safety talk each day to remind everyone of potential dangers, proper techniques, and safe use of new equipment for upcoming work. It's also useful to go over any "near misses" from the previous day.

*A deconstruction site is always a potentially dangerous location, and safety should always come first.*

## The Safety Spotter

While it's everyone's job to be safety conscious, it's a good idea to have a designated safety person, or "spotter." Ideally, this person would have formal job-site safety training or enough experience to identify hazards and know what to do about them. The safety person's job starts before any material removal begins: identify potential hazards, erect barricades or danger tape around hazardous areas, and make sure there is the proper safety equipment on site for the project. In addition to reminding workers to wear their safety glasses, earplugs, dust mask, gloves, and hard hat, this person needs to constantly monitor the changing conditions of the site and take action to correct situations that might create a hazard.

## Make Safety a Priority

Several factors affect safety and environmental health on a project, including the condition of the building and job site, the tools you use, weather conditions, the cleanliness of the job site, and the presence of materials coated with lead-based paint (LBP) and other hazardous substances. We discuss these factors and the safety equipment you'll need to maximize your safety. We also present some commonsense steps you can take if you do have to deal with an injury or other emergency.

## Building and job-site conditions

As part of your initial assessment of the project, you should note any problems with the building or site that might affect your safety and health. It's time well spent to make corrections to existing unsafe conditions at the beginning of a project. Areas that contain rotted members, bowed or buckled framing, or other structural problems should be cordoned off, shored up, or otherwise corrected. Plywood or boards should be nailed over any holes in the flooring; and any missing railings on stairs, balconies, or other places at height should be similarly addressed. The presence of any hazardous materials should be indicated until remediation can take place. Areas outside the building that pose a risk should be remedied, barricaded, or flagged. This is the job of the project supervisor or the designated safety person.

*Holes in floors and ceilings add to the dangers on a deconstruction site.*

## Some Job-Site Hazards

- Unstable parts of a building, including masonry, rotted wood members, holes in the floor, unprotected openings at windows and stairs, broken railings

- Holes in the ground, uneven ground, or a steep slope

- The presence of roots, pipes, posts, barbed-wire fence, overhead wires, branches overhanging the roof, general debris, and insect nests

- Biological hazards such as animal droppings, carcasses, human waste, mold, flaking lead-based paint, loose or other flaking debris that can cause dusty conditions

- Broken window panes or broken glass underfoot

## Tools

The tools you use should be up to the tasks demanded of them, maintained in good condition, and used the way they were designed to be used. Although each person will want to periodically inspect the condition of the tools in his or her tool belt, everyone should keep an eye out for problems with job-site tools, such as extension cords or corded power tools with frayed insulation, missing or broken parts or blade guards, or other problems. If you find a tool with a problem, don't just set it aside. Another person may pick it up not knowing its unsafe condition. Tag it, flag it, or give it to the safety person so no one uses it until it has been repaired.

## Weather conditions

As you take apart a building, you progressively expose your site to the vagaries of the outside environment. Rain, snow, and dew can create slippery conditions, meaning you need to be extra careful. Wet conditions also increase the risk of electrical shocks, so we recommend you install GFCI (ground-fault circuit interrupt) protection at the job-site electrical panel to safeguard all downstream circuits. If you use a generator, follow the manufacturer's recommendations for GFCI protection and grounding.

Exposure to extremely hot and/or humid conditions can lead to heat exhaustion and heatstroke. In the warmer months, make sure everybody drinks plenty of water, spends as much time in the shade as possible, and isn't overtaxed to the point that he or she might make serious mistakes. At the other extreme, cold or snowy weather has its own difficulties. If conditions are cold, it's always prudent to have a place to warm up. Frequent breaks for warm liquids and to warm up hands make for safer work.

*Wet weather* can make for slick conditions, which are just as dangerous whether deconstructing an industrial building with heavy equipment or unbuilding a house by hand.

*Daily cleanup makes the site safer and more pleasant to work in.*

## Cleanliness

A clean job site is a safer job site. The risk of tripping, slipping, or stepping on a nail decreases if debris is regularly removed and not allowed to pile up in the work areas. Also, a clean workplace can have a positive psychological effect. Most of us enjoy our work more if we start the day in a clean space. You might designate the last half hour of the workday as cleanup time to accomplish this.

Because of the sheer number of nails on an unbuilding site, a typical cause of injury is a nail through a boot or a trip and fall when salvaged wood members are piled up or set in places where people walk. *Always* place wood out of traffic areas with the nails facing down.

Assuming you've identified the potential hazards on site, your last line of defense for safety is to use personal safety gear. Protecting the most vulnerable parts of your body—eyes, ears, hands, feet, lungs, and head—is easily achieved. The following items are essential for any deconstruction work.

## Safety glasses

The single most important item in your toolbox may be your safety glasses. We prefer wraparound safety glass to goggles, which can easily fog and obscure your vision. Safety glasses, even the fashionable ones, are inexpensive. Buy an extra pair for your toolbox.

*Wraparound safety glasses are more comfortable than goggles and don't fog as easily.*

*Wear clothing appropriate for the work; good boots and a hard hat will reduce injury on the job site.*

## Gloves

Gloves help prevent blisters, splinters, and cuts. The best protection is with leather or a tough synthetic—cotton gloves are for the garden. Disposable rubber gloves are useful if removing insecticides, solvents, or other chemical containers.

## Ear protection

Always wear ear protection when loud power tools and generators are in use. Disposable earplugs are inexpensive, especially if you buy them by the box; keep a box handy on the job site.

## Work boots

Even if you're just doing simple soft-stripping, wear a good pair of work boots that provide traction and ankle support. Steel-toed boots are even better and provide added protection for your toes. Boots with puncture-resistant soles can save you from injury caused by stepping on a nail.

## Hard hat

Wear a hard hat whenever there's a risk of objects falling from above or a chance of getting smacked in the head by someone carrying a piece of lumber. On some job sites, hard hats are required at all times; we think that's a good idea.

## Long pants/long sleeves

Long pants may be hotter than shorts, but they provide a valuable layer of protection. Double-fronted pants, such as those made by Carhartt®, provide even more protection and added durability on the front of the legs. A heavy-duty long-sleeved shirt can help prevent scratches on your arms.

## Dust masks/respirators

A simple disposable respirator (or dust mask) provides good protection against dust, fibers (such as from insulation), and other particulates

*It's important to wear gloves that are tough enough to protect hands from splinters, sharp metal edges, and naily lumber.*

*A variety of respirators* are available for exposure to different materials. Visit a safety supply store for advice on appropriate types and proper usage.

and should suffice for limited use. The disposable respirators you'll find at your building-supply store have two ratings: N-99 and N-95, which are the particulate efficiency ratings (99 percent and 95 percent, respectively). Buy the best you can afford. However, if you plan to do this kind of work more intensively and produce dust from LBP or other nonoil hazardous materials, you should use a half-mask respirator with N-100 filters. It's important to get a half-mask respirator fitted properly to your face to prevent leakage, so buy from a supplier who can help you with fitting and care. Look under "safety equipment" in the yellow pages.

*A disposable respirator* (rated N-95 or N-99) works well for most situations on the deconstruction site (here, vacuuming cellulose insulation from a wall). The mask should fit tight around the nose.

*If you produce dust* from lead-based paint, asbestos, or other hazardous materials, invest in a half-mask respirator with N-100 filters and have it properly fitted to your face.

## Working at Height

Working off the ground, whether on a ladder removing fixtures or up on a roof stripping shingles, is inherently dangerous. Falling, of course, is the biggest hazard; and usually the greater the height, the greater the injury. However, there is truth to the adage that "it's not the fall that kills you, it's what you land on." So, falling on a nail from the first rung of a ladder could be worse than falling off a scaffold and landing on soft ground. Another reason to keep the site clean.

## Ladders

Buy the highest-quality ladder you can afford and make sure you read the manufacturer's warning labels on the ladder (there are many!). Even the best ladder is only as good as the person using it—there's always the possibility of injury if you don't set it up properly or use it beyond its capabilities. Get up on a ladder only if the feet are on firm, level ground and all brackets securing the legs together are locked in place. Never climb a ladder you didn't set up without first checking that it's positioned correctly (someone may have leaned the ladder against a building without intending to climb it). Also, always face the ladder when using it. Standing on the steps with your back to the ladder is dangerous.

*An extension ladder* should be set up on firm, level footing and leaned at the proper angle.

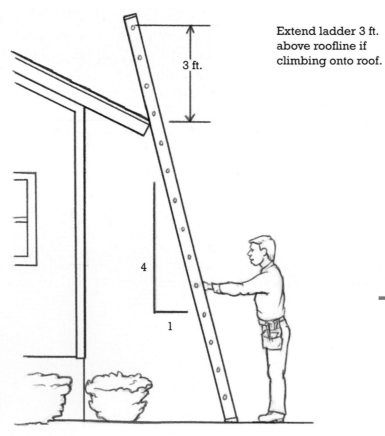

Extend ladder 3 ft. above roofline if climbing onto roof.

3 ft.

4

1

*A good rule* of thumb for positioning a ladder is to stand with your toes at the base of the ladder with your arms fully extended. If your arms are bent, the ladder is too steep. If you can't reach the ladder, then the ladder is at too shallow an angle.

If you are trying to access a roof with an extension ladder, the top rung should extend at least 3 ft. above the level at which you'll step off the ladder. This is to ensure that you have a handhold when stepping onto the upper level. If possible, it's also a good idea to tie off the ladder at the eaves with a rope for added security. When placing an extension ladder against a building, there is an optimal angle at which to lean it. The U.S. Occupational Safety and Health Administration (OSHA) suggests a 4:1 ratio—that is, place the feet of the ladder 1 ft. away from the wall for every 4 ft. of ladder height. For example, if you have a 16-ft. ladder fully extended against the side of a house, the base should be about 4 ft. from the wall.

Placing a ladder in a doorway or other high-traffic area increases the risk of injury. If a ladder must be used near a lockable door, lock it. If that isn't possible, at least inform others working in the area so they don't inadvertently knock you over with an armload of 2×4s.

## Fall protection

One of the most common OSHA violations on a construction site is inadequate fall protection. According to OSHA, you are "working at height" any time you're standing 6 ft. or more above the level below. Any two-story building or the roof of a single-story building will put you in this situation as will standing on a floor with a hole in it that's big enough to fall through. A deconstruction site is unique in that it involves removing the building elements from around you to create an ever-moving "leading edge" from which you can fall. When you're removing a roof or second floor, you're actually creating more openings to fall through. A first step

*A fall-protection harness makes working at height safer and more comfortable.*

in preventing falls is to make sure that anybody who might work at heights is fit for it. Working at heights requires good physical condition as well as a disposition that's comfortable with heights.

There are several aspects to fall protection to consider, including controlling and monitoring access to areas where there's a risk of falling, erecting temporary railings and barricades as you are removing the building structure, and using fall-protection harnesses.

As your work progresses through a building, you will end up working into the areas that were flagged as dangerous (holes in the floor, rotted supports, and so on). Where you might

earlier have put up warning flags to keep people away, you may now need to erect barricades to prevent workers active in the area from falling. If the holes are small enough, you can nail down plywood to cover them, though you need to check how far the rot or other damage extends to make sure the patch is sound.

As you remove materials from the building, you'll actually be creating fall hazards—for example when cutting a window opening down to the floor so you can shovel debris into a trash container below. Be sure to add a guardrail for any opening you create. Generally, scaffolding, floor openings, ramps, and open areas in which someone can fall from one level to another need to be protected by a guardrail, which according to OSHA consists of a top rail, a midrail, and a toe board.

When you have to work outside the limits of barricades, guardrails, and the safety of a level surface, you'll need a fall-protection harness. A cousin to a mountain climber's harness, a fall-protection harness wraps around your body and tethers you high on your back. The harness isn't the only component of this safety system. You need a lanyard to connect you to a rope or cable that is in turn secured to a beam or anchor. The lanyard acts like a short Bungee® cord to soften the impact of being stopped by the rope or cable, and the rope or cable gives you some room to move around the anchor point.

*The lanyard that tethers you to your safety harness can be secured directly to a solid object, such as the large beam shown above; but you can also use an intermediate rope or cable that is in turn secured to a beam or other anchor point. Remember that the lanyard must be kept short enough to keep you from hitting the ground if you fall.*

*On this roof, two large brackets with a metal ring were securely fastened to the roof framing at each end of the roof. A steel cable was stretched between the rings and served as a convenient location to clip on lanyards as workers moved across the roof surface.*

## First Aid and Medical Services

Before work starts, think about how you will deal with injuries, both minor and serious. Minor injuries such as splinters, small cuts and abrasions, bruised fingers, and sprained ankles are not uncommon on a deconstruction site. Maintaining a well-stocked first-aid kit is a good first step in dealing with such injuries. It's a good idea also to have an ice pack on hand for joint injuries. A source of clean water should be readily available in case someone needs to flush his or her eyes of dust or dirt.

More serious injuries usually require outside help, and you should know where the local hospital is and how to get there. First, make sure your cell phone works at the site (there are still dead spots in some areas) and is charged at all times. It's a good idea to have more than one cell phone on site because if the only cell phone leaves the job with a person fetching supplies or lunch, the rest of the crew may be without communication if an emergency arises. Also, if you are working on a site that could be difficult to find, as part of your site planning you will want to print instructions for the most direct route to your site from the nearest hospital or clinic. This is especially important if you have workers on site who are not familiar with the area and need to give instructions to emergency personnel. Along with these directions, post the direct phone numbers for local police, ambulance service, and fire department in a conspicuous place. Make sure all workers know where to find this information.

Finally, we encourage anyone who's planning on unbuilding to take a first-aid course, including cardiopulmonary resuscitation (CPR). The American Red Cross has a wide range of courses to fit most people's schedules and budget.

## Fire Prevention and Protection

Working on an older building with lots of dry lumber and disconnected water service means that any kind of fire can quickly get out of control. Most fires can be prevented by using a little common sense. When possible, eliminate piles of combustibles, such as scrap wood, trash, and paper. When you can't, make sure such materials are isolated and out of reach of the building in case they do catch fire. Keep any liquid fuels out of the building and in a designated, nonflammable area. Obviously, you don't want to set up your generator (or other gas-driven tool) and refueling area on a bed of dry weeds. Be mindful of hot exhaust pipes (from cars, heavy equipment, and other gas-driven equipment) and grassy areas that might catch fire. Don't allow smoking in the building (designate a specific area—well away from the building and materials—for any smokers).

### Fire Extinguishers

You should always have at least one, and preferably several, fire extinguishers on site. Locate one on each floor and one where you store fuels. Fire extinguishers are rated for different types of fires; the designation "ABC" covers all types of fires (fuels, electrical, and other combustibles) and should be adequate.

## Lead-Based Paint Hazards

You don't have to deconstruct many buildings before you run into lead-based paint. Any building built before 1978 should be tested for the presence of lead. Lead in your building affects both the care you should exercise while deconstructing the building as well as what you do with the materials you harvest. And paint isn't the only culprit in a house. Some old varnishes (typically from pre-1930s houses) contained lead, which means that varnish on wood trim, doors, hardwood floors (if not already resanded), stair treads, and so on, can contain lead, though typically at lower levels than paint.

In most cases, knowing if lead is present is more important than knowing the exact concentration of lead. Several test methods are available. Probably the most common, and least expensive, is a chemical spot test, which determines presence of the metal, not its concentration. Household lead test kits are available from hardware stores or online.

*The applicator* contains a vial enclosed in a cardboard tube. When broken, the chemicals inside mix together. The mixture is squeezed out and swabbed onto the paint; if lead is present, there is a reaction, which causes a change in color.

*Here, the red coloration* indicates a positive test for the presence of lead.

*Easy-to-use* lead test kits are widely available. This kit, purchased at a hardware store, contains two applicators.

## Cutting Paint to Find Lead

Because a house may have been painted several times, a layer (or layers) of lead-based paint may have a topcoat of nonlead paint. In this case, a test swab applied to the outermost layer may not indicate the presence of the lead. Use a utility knife to cut away a section of the wood and test along the paint line.

*An initial test on the top layer did not indicate lead. However, cutting through all paint layers exposed an old lead-based paint layer that tested positive.*

## Asbestos Hazards

First and foremost, we recommend you leave asbestos detection and removal to the professionals, unless you want to go through the training and get licensed. We wouldn't start any deconstruction project until the building had been inspected and any known asbestos remediated.

It's a good idea to educate yourself about the common places you might find asbestos because it could be missed in an inspection or lie buried under layers of materials. If in any doubt about a suspect material, have it tested. We provide some examples of where asbestos can be found in "Asbestos Hazards in the Home" on p. 146, but bear in mind that the list is by no means exhaustive.

*Working with asbestos requires training and licensing. These workers are suiting up for remediation of a military building.*

*Transite siding is an asbestos-based cement siding found in military buildings and homes constructed in the 1940s and 1950s. Considered nonfriable if unbroken, exposure to transite's asbestos fibers can be minimized if it is removed carefully. If you run into transite, you should check with local ordinances regarding requirements on removal and disposal.*

*Vermiculite insulation* can be found in old homes. Some, but not all, brands contain asbestos; if you find vermiculite, bag it and have it tested.

*The covering* on some old electrical wire can contain asbestos. These covers are generally nonfriable and safe to handle as long as you don't fray them.

## Protecting yourself and the environment

Both lead and asbestos are dangerous if inhaled, so your best defense is to wear a disposable or half-mask respirator whenever dust is generated. Exposure to both hazards can be minimized if you take some simple precautions. Most important: When possible, avoid sanding, grinding, and sawing anything that's either coated with lead paint or could contain friable asbestos. (If you do ever have to sand, grind, or saw wood with LBP, do it outside, wash your hands afterward, and change your clothes so you don't contaminate the people you come in contact with.)

If removing interior trim with intact paint, the pieces of paint that fall off while prying off the wood will likely be large and so won't travel very far. A disposable respirator should stop them. However, if you're pounding plaster off a wall and lead

## Asbestos Hazards in the Home

- Roofing and siding shingles can be made of asbestos cement (Transite), as can the mastic used to seal roof penetrations around vents, the caulking in windows, and the mastics used to glue down flooring.

- Asbestos insulation may be present in houses built between 1930 and 1950.

- Asbestos may be present in textured paint and patching compounds used on wall and ceiling joints. Its use was banned in 1977.

- Artificial ashes and embers sold for use in gas-fired fireplaces may contain asbestos.

- Older products such as stovetop pads may have some asbestos compounds.

- Walls and floors around wood-burning stoves may be protected with asbestos paper, millboard, or cement sheets.

- Asbestos is found in some vinyl floor tiles (usually 9 in. by 9 in.) and the backing on vinyl sheet flooring and adhesives.

- Hot-water and steam pipes in older houses may be coated with an asbestos material or covered with asbestos-based insulation.

- The covering on some old electrical wire may contain asbestos.

- Oil and coal furnaces and door gaskets may have asbestos insulation.

- Asbestos may be present in soundproofing or decorative material sprayed on walls and ceilings.

is in any of the paint layers, you will produce lead dust. In that case, it's best to wear a fitted half-mask respirator with an N-100 filter. To further protect yourself, you'll want to use work gloves, though you'll still want to wash your hands before eating lunch and leaving work.

For the general health of your lungs, it's a good idea to get as much ventilation through the building as possible. Open windows and use a fan to vent any dust you generate. If the dust is particularly prevalent on the floor or walls, use a HEPA-filter vacuum to remove as much as you can as you are working.

If flaking paint is an issue, such as on siding or trim, avoid breaking flakes off so they don't end up all over the house, yard, and tracked into your truck and back to your home. If flaking is serious, put a tarp down along the house while removing the siding to catch all the paint flakes. You also want to scrape off the loose flakes as you go: This helps prevent poisoning the soil around the house and potential lead exposure to the new landowner who might unknowingly plant a vegetable

garden or allow kids to play in a contaminated spot. Collect the flakes in a bucket or trash bag and dispose of according to local ordinance.

Another simple way to keep the dust down is to use a garden sprayer and lightly mist surfaces, such as plaster and drywall, that have lead paint on them. The water helps bind the particles to the underlying material so it won't float around in the room and be inhaled.

*When dealing with lead-based paint, a Tyvek suit can keep lead dust off your clothes and prevent tracking it back to your car and home.*

*Max Taubert (left) and Pete Krieger (right) of the Duluth Timber Company have reclaimed and distributed for reuse millions of board feet of big timber.*

"Maybe it's the boy in me,
but I love hearing the big crash as
parts of buildings go down."

# Big Timber: Logging the Industrial Forest

**Max Taubert**
**Peter Krieger**

Max Taubert loves to dismantle big timber buildings. So much so, that he founded what is now one of the largest distributors of salvaged timber in North America, the Duluth Timber Company, headquartered in Duluth, Minnesota. Max estimates he's salvaged or bought and sold about 40 million bd. ft. of timber in the last 20 years.

Max got his first taste of salvage in the 1970s, when he deconstructed a large brick house in Painesville, Ohio, that was to be torn down for a redevelopment project. A feed mill and a few barns followed and, Max honed his deconstruction skills. "For a time during the 1970s I also worked as a carpenter's helper. I handled a lot of lumber, and I was seeing the increased use of second-growth and plantation-grown framing lumber. It was clear to me that the lumber I was finding on my salvage jobs was of much higher quality."

Some years later, Max got into timber salvage in a big way when he landed the deconstruction contract on a giant grain elevator in Superior, Wisconsin. A wooden trestle and steel drawbridge followed. "Progressively, I got serious about bigger and bigger

structures," says Max, "Maybe it's the boy in me, but I love hearing the big crash as parts of buildings go down."

By the early 1980s the market for reclaimed lumber started to gain momentum. Max stopped dismantling and started buying, regionally and then nationally, from other demolition contractors. In 1990, Max met the man who is now his general manager, Peter Krieger.

Peter's path to working with reclaimed timbers began as a daydream in Minnesota's Twin Cities' suburbia. "I was working as an overeducated, underskilled carpenter for an architect's design/build firm. We were building houses wrapped in plastic and trimmed in plastic so the people who bought them could fill them up with plastic. We used to talk about going to a timber-frame workshop, just to experience the craftsmanship of handbuilding something to last." That fateful workshop inspired Peter to get into timber framing and he landed a job with the G. R. Plume Company, which is where he was working when Max found him. For the next 5 years, Peter found himself working exclusively with Max's big timbers. It was technically

challenging work, through which he got to know every knothole and flaw. "I came to appreciate the huge amount of history embedded in each piece of wood," he says.

"For example, take a redwood wine tank, which Max and I have salvaged a lot of. The tree was logged in the 1920s, it was milled and installed in a winery, where it worked for, oh, 70 years. Then we disassemble it by hand, stack it carefully, and remill it using what is essentially the same technology as in 1920. And then skilled craftspeople reinstall it in its new life. We're honoring not only the tree and the resource but all the human labor that went into it."

Today, Max and Peter still buy and sell big timbers, of course, but also vast quantities of smaller stock (including the redwood and cypress from wine, water, and pickle tanks) and some oddball stuff like bleacher seats and gymnasium flooring. "It's satisfying when you get to match up one person's abandoned material with another's newly discovered need," Peter notes.

# SITE PREPARATION AND SOFT-STRIPPING

## Chapter 6

*In theory there is no difference between theory and practice. In practice there is.*

**—Yogi Berra**

**I**n the last five chapters we've talked in general terms about unbuilding and materials salvage as well as planning and organizing a deconstruction project. We've focused on the important questions to be asked before getting started and on considerations of logistics, site evaluation, and safety. In this chapter and the next, we'll get out of the armchair and put all this theory to work on an actual unbuilding project—deconstructing a single-family bungalow home. As we step through the actual deconstruction process of this house, we offer tips and advice that are useful for this and other situations you may run across.

We were invited to work alongside the building materials reuse company contracted to deconstruct this house, The ReUse People of America (TRP). Based in Oakland, California, and in business since 1993, TRP has deconstructed more than 1,000 houses. It has developed an extensive network of trained contractors and highly organized material inventory and delivery systems. Also onsite with TRP was one of its licensed deconstruction contractors, St. Helena Construction, located in St. Helena, California.

*This California bungalow was built in 1901 and had been extensively remodeled over the years. The owner wanted to deconstruct the house both to make way for construction of a new home and to salvage the materials.*

As we have discussed, many factors influence the exact sequence of events on a deconstruction site. Building type, geography, site constraints, and weather all have to be considered in determining the succession of tasks to take the building to the ground. Also, the value of materials saved from a house may not be the same in all markets, and no single house can illustrate all conditions. This house was in good condition (with little rot or other damage) and sat on a flat lot in warm, dry California with little chance of rain during the scheduled deconstruction. Although not every site is this ideal, we will point out ways to handle other climates, sites, and conditions as appropriate.

First, we discuss some specifics of this house and then guide you through the soft-stripping (or nonstructural) phase, detailing the removal of interior items, including cabinets, light fixtures, doors, windows, flooring, and other reusable items. The next chapter focuses on deconstructing the remainder of the house, highlighting the removal of structural materials (roof, joists, and rafters), plaster, and drywall and the salvage of siding.

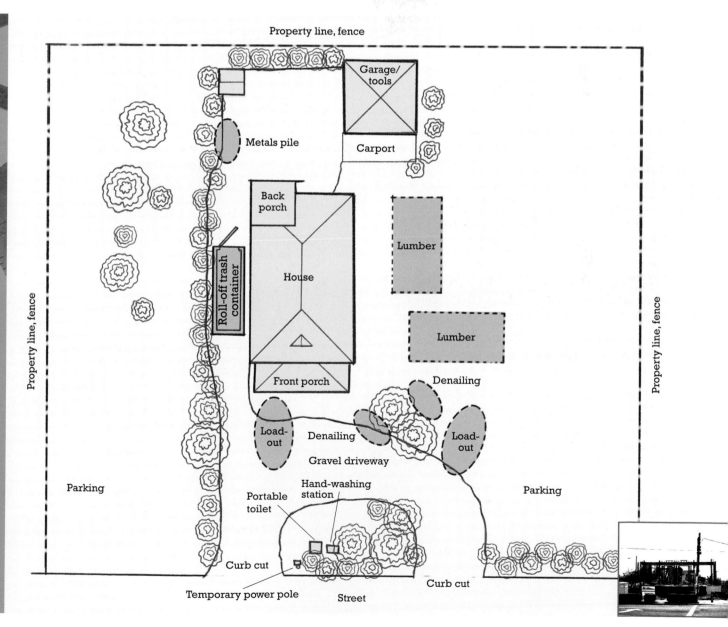

SITE PLAN

Property line, fence

Garage/tools

Metals pile

Carport

Back porch

Lumber

Roll-off trash container

House

Lumber

Denailing

Front porch

Load-out

Denailing

Load-out

Gravel driveway

Parking

Parking

Hand-washing station

Portable toilet

Curb cut

Temporary power pole

Street

Curb cut

Property line, fence

Property line, fence

## House and Site Characteristics

The project documented here was a story-and-a-half bungalow in northern California built in 1901. This modest-size three-bedroom house, located on a relatively large lot (nearly an acre), is typical of many teardowns, in which a smaller outdated home is removed to accommodate a larger new home.

The house was originally one story, but a stairway and second-story bedroom had been added (you can see the shed dormer addition in the photos). Although this home possessed some original architectural features, it was not considered historic by local or national standards. It appeared to have been remodeled at least twice, and as a result a mixture of materials was present, including lath and plaster on the first floor and drywall on the second. Three types of insulation material were found: blown-in cellulose, vermiculite, and fiberglass. The original parts

*Except for some bushes* and an overhanging tree, access to all four sides of the house was fairly open. On the side where the roll-off trash container would go was a large shed dormer (top photo below); the other side of the house was completely open (below left); out back (bottom photo below) was the garage and an enclosed deck on the rear of the house.

*A large built-in* storage unit, which appeared to be original to the house, separated the dining room from the kitchen. The open layout of the kitchen made cabinet removal easy.

of the house were built using balloon framing, and the materials and components fit the period and region: redwood siding, Douglas fir flooring, and single-pane windows. A few windows had the original wood frames with individual lights, though most were replacement wood or aluminum framed.

The lot itself was fairly level, but there were many trees, shrubs, and ornamental plants that we needed to work around. Two nests of California quail eggs were discovered in the bushes next to the house, and we took great care to minimize disturbance to our avian friends. There were two outbuildings on the site, a garage and a small shed, neither of which got in the way of the deconstruction process. The lockable garage was nearly full of the owner's belongings, but some space was made available for our tool storage.

Before starting the actual deconstruction, all utilities were shut off, including gas and water. Overhead electrical service to the house was disconnected and a temporary construction pole, identical to that used on new construction sites, was installed near the street. A heavy-duty electrical cord was run from this street panel to a portable job-site electrical box.

For safety reasons, orange plastic construction fencing was erected along the borders of the lot where there was no existing lot-line fence; a section of fencing was also used to close off the driveway at the end of the day. We put up a sign to identify the activity on site. Before work began, the site supervisor walked around the site to identify existing hazards or any other safety concerns. An open bucket of what appeared to be motor oil was found in the back shed and required disposal at the local hazardous waste facility.

*Before getting started,* all the utilities were cut off, including the water main from the city. Here, a curb key is used to shut off the water meter at the street.

*This portable job-site* electrical outlet box was connected to the service pole electrical panel with a long, heavy-duty electrical cord. The box contains a number of three-prong outlets and a circuit breaker for each one.

*A temporary job-site* electrical pole was erected so all electricity in the house could be disconnected.

*Running snow fencing* along the borders of the lot helped secure the job site.

We identified the best spot for the roll-off trash container and other work areas before work began. A 40-yd. container was located as close to the house as possible and under both a first- and second-story window to make loading from the roof and from the house interior easy. The side we chose had the advantage of an existing gravel driveway. Though some trees crowded the container dropoff on the east side of the house, only a couple of smaller limbs needed to be trimmed.

Locating the trash container just outside the windows made it easy to pitch debris directly into the bin once the window was taken out. A specific spot was also chosen for recyclables. All metals and appliances were piled in a single, out-of-the-way location for later pickup by a metal recycler (see "Site Plan" on p. 151). Because of the rural location of this project, recycling clean wood scrap was not practical. The closest wood recycler was a 2-hour drive away and the cost of transportation for recycling far exceeded the cost of disposal. On a project with a greater volume of clean

*A job-site sign* can help others locate your site.

*The 40-yd. trash container* arrived early on the first morning. Here, a worker directs the roll-off driver down the narrow gravel drive so the container can be placed under a large window.

*The roll-off* is in place. (The taped-off area in front of the house is to protect a nesting quail family.)

*Metals to be recycled,* including aluminum window frames, stainless-steel sinks, and appliances, were kept together in a growing pile, well out of the way of the action.

*A heavy-duty locking storage box held most of the hand and power tools for the job. It was locked and then placed in the locked garage each night.*

wood and closer to a wood recycler, disposal costs could have been lowered by recycling this scrap wood.

A denailing station was tentatively located off the front west side of the house under a shade tree midway between the house and the lumber storage area. The area for stacking lumber was located for easy truck access on the gravel parking area in front of the house. A Porta Potti® and hand-washing station were located away from the house but just off the driveway for easy delivery and pickup. A lockable job-site toolbox

was located in the garage of the house, and all tools were returned to this location at the end of each workday.

Before starting the soft-stripping process, we had to attend to a few other items. Vermiculite insulation was found in the attic and had to be tested because this type of insulation (depending on where mined) can contain asbestos. Fortunately, the test result was negative for asbestos so no special disposal procedures were necessary.

As with any construction job, it is important to hold regular safety meetings. On this project, a comprehensive orientation and safety meeting was held on the first morning of work to familiarize everyone with

**HERE'S A TIP**

### Dealing with Roll-Offs

Trash containers aren't pretty and aren't sweet-smelling. What they are is expensive. Trash is money, at least you should look at it that way, as you will be paying to have the container delivered and then picked up, paying for its stay at your site, and paying to have it dumped at the local landfill. While you can't do much about the costs of the delivery, rent, and local tipping fees—because they are usually nonnegotiable—you can pack the debris tightly in the container to maximize the amount of material in each load. That will help minimize *your* costs.

Try to lay longer items all in one direction instead of randomly crisscrossed. You'll fit more in each container and minimize pickups and deliveries. It's usually more cost-effective to use a 40-yd. container, but the downside is that they're considerably taller than the 20-yd. versions and you can't see over the sides when you're standing next to one. You'll have to decide what makes the most sense for your job site.

*Two trash containers were used at this deconstruction site. The one on the left was just for smaller debris and the one on the right for longer boards that were too damaged for recovery. Because care was taken to tightly stack the lumber, all of it fit into one roll-off, avoiding the cost of another container delivery.*

*Vermiculite insulation* was found in the attic. A sample was bagged, tagged, and sent to a lab for analysis, but fortunately it came back negative for asbestos.

*The project manager* and designated "safety person" held a safety meeting at the beginning of the project and periodically thereafter.

the work plan, special site conditions, hazards, and safety issues. A brief meeting was held each morning to review any logistical, safety, or planning issues.

## Soft-Stripping

The first phase of unbuilding this house involved soft-stripping reusable light and fan fixtures, kitchen cabinets, built-ins, doors, windows, porch railing and bracketry, plumbing fixtures, and flooring. Before any material was removed, the TRP project manager surveyed the site and inventoried everything. All materials to be resold or recycled were labeled with a color-coded identification sticker. By using a numbering system, each item could be tracked, which is especially important for inventory purposes.

*Adhesive tags* placed on each reusable item ensured accurate tracking of resale inventory.

*A green adhesive tag* was affixed to everything that would later be recycled. In this case, a single-glazed aluminum window, which had no salvage value, was tagged for the metal recyclables pile.

## Light fixtures

Though none of the light fixtures in the house was original, there was still enough value in them to salvage. Several ceiling fans were also salvaged. Fragile items of this type require special handling to avoid breakage, especially if they have glass globes. The larger fixtures and fans were packed in cardboard boxes; smaller fixtures fit in 5-gal. plastic pails.

All of the original light switches in the house had been replaced with newer switches and were not worth saving. However, in some older homes, the original two-button light switches and switch plates are worth salvaging because of their value to people doing historic restorations.

*Although they* were neither original nor antique, the light fixtures and ceiling fans had resale value at The ReUse People's retail warehouse.

## Kitchen cabinets

Once all the light fixtures were removed, we moved on to the kitchen cabinets. It's usually easiest to start with the upper cabinets, but it is a matter of preference. In newer homes, cabinets typically are held to the studs with several long screws. This is where a cordless screwdriver or drill comes in handy.

*Cabinets are awkward to handle, and extra help is welcome. In this case, because the cabinets in question are uppers, three people worked together to carefully remove the units, which were lag-bolted to the wall—an example of the typical surprises that await anyone who deconstructs an old house.*

In this house, for some unknown reason (and old houses are full of such oddities), the cabinets were attached with lag bolts. They did take a bit longer to remove because we had to use a ratchet wrench. In many older homes, cabinets are built in—that is, they are nailed together and to the walls, making removal and reuse more difficult. It's always a good idea to photograph the kitchen before removing the cabinets. It will help

a potential customer visualize how the cabinets will look in his or her project and can serve as a good sales tool.

Plan on having at least one helper to remove cabinets because the units can be heavy. Also, it usually takes one person to support the unit while the other unscrews it.

An advantage of newer, modular cabinets is that they come in standard sizes. So, along with your photos of the kitchen, make a list of the cabinets and note whether they are uppers or lowers, record their width, and write a description (for example, number of drawers and swing direction). Lower cabinets are usually a standard height of 34 in., but uppers can vary in height, so list that dimension as well.

*Modular cabinets are easy to remove, transport, and integrate into a new kitchen.*

*Once the sink is out, the lower sink cabinet comes out in the same manner as the others.*

We removed all the uppers first and then moved on to the lowers. The procedure is the same, though you will usually have to remove the toekick before you can separate the individual lowers. The most difficult cabinet to remove is typically the sink unit. You have to crawl under the sink and cut both the water supply and waste lines (remember there is dirty water in the trap!), as well as loosen or remove the clips that hold down the sink. Depending on sink type, caulking may be the only thing holding it into place. In that case, cut through the caulking with a utility knife and the sink should pop up. To minimize the time you are on your back under the sink, remove the faucet *after* the sink has been pulled out. Again, use a plastic bag to keep all parts together for easier reinstallation.

*Here's the kitchen* once all the cabinets are out.

*A modern kitchen* faucet comes with about a dozen individual pieces in various sizes, most of which are fairly hard to replace and which, if lost, seriously lower the value of the faucet. Again, keeping all the pieces in a sealable plastic bag and taping the bag to the faucet ensures that all will stay together.

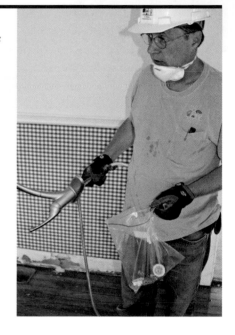

*Soft-stripped items* were temporarily stored in the large living room space for easy transporting to the truck, which would back up to the front porch. Make sure that when storing salvaged materials you leave a wide enough path for traffic flow. Plenty of room allows everyone to get by with tools and boards without running the risk of tripping and injury.

The lighting and cabinets were temporarily stored in the living room until the soft-stripping was finished and a truck could be scheduled to transport the items to the reuse warehouse.

## Built-ins

Built-in cabinets can be difficult to remove. While the upper and lower kitchen cabinets in this house were installed with lag bolts and were easy enough to remove, a cabinet built into the wall between the kitchen and the dining room was more challenging. This painted cabinet, original to the house, had a passthrough to move food from the kitchen to the dining table. It was questionable whether the resale value of the unit made it worth salvaging, but it needed to be removed and we decided to spend a little extra time to take it out of the wall carefully. Because the plaster finish was applied right up to its edge and we suspected the cabinet was nailed into the surrounding wall studs,

we used a hammer to break out the plaster and gain access to the nails. We then used a reciprocating saw to cut the nails around the unit, freeing it.

## Opening up the house

Moving some of the bulkier materials out of a house can be a chore, especially trying to squeeze cabinets through existing door frames. If possible (it may not be practical if you're just doing a soft-strip), opening up part of the house to more easily move materials out can save time and energy. As shown in the photos on p. 164, we removed the front windows and the wall around the windows to allow for a clear exit path. Of course, we didn't make this decision without first determining that this wall would carry the load of the second floor and roof above. We inspected the wall and determined that it carried only half the porch weight and little else because it was the gable end of the

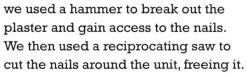

*A built-in cabinet* between the kitchen and the dining room was cut out and removed. The trim was removed first, then the plaster around the built-in was hammered out. After cutting around the whole unit with a long metal-cutting blade in a reciprocating saw, the unit was free and could be tilted out. The cabinet wasn't particularly heavy, but it was large and bulky; thus to preserve it in as intact a condition as possible, the job required two people.

**Load-Bearing Walls**

We had a licensed building contractor overseeing this project, so we felt comfortable removing the front window wall without fear that anything would collapse. But if you have any doubt about what is and is not a load-bearing wall, make sure to consult a competent builder before wielding your saw and hammer.

house. It appeared strong enough and proved so, as it did not deflect at all when the window and adjoining wall were removed.

Removing the built-in cabinet between the kitchen and the dining room opened up a large space for material transfer from the kitchen through the dining room and onto the porch. This open area not only made it easier to move materials out of the house but also provided air and natural light, which made the subsequent work safer and more pleasant.

*To allow for the easy removal and transfer of building materials, fixtures, cabinets, and even a fiberglass shower unit, a large bay window was removed and the opening onto the front porch widened. First, the siding was taken off to allow access to the framing, and then the connections were cut with a reciprocating saw to allow the wall section to be pulled out. It was determined before taking out the wall that the building would remain structurally stable.*

*Taking out the built-in* opened up a wide space that made it easier to carry materials out of the house.

### A Prybar for Thin Trim

At times, a flat prybar is too thick and will do damage if you try to hammer behind delicate trim pieces. The thinner trim bar shown below works well in these cases. By lightly tapping the flat end of the bar behind the trim and using a little prying action, the trim will usually pop out from the wall far enough that the flat prybar can be slipped behind. The flat prybar provides enough leverage to then pry the trim out from the wall.

*A small, thin prybar is indispensable for removing delicate trim.*

## Windows and doors

We removed the windows and doors next, using a similar procedure for both. The house had a mixture of window types, including fixed glass, casement, double-hung, and sliding. Only a few windows were original; replacement windows were double-hung or aluminum single pane. The aluminum was recycled because the single-pane windows had little or no resale value.

### REMOVING THE WINDOW TRIM

The first step in removing a window is to shut and latch the window. This will prevent racking and damage to the unit as you are taking it out. The next step is to remove the trim. Here, we show the exterior trim being removed first, but the order is not critical (just be aware that you may need a ladder or other platform to access the exterior trim). Keep in mind that the goal is to remove each window as a unit so it can be prehung in a new opening.

Many times, you'll find that the exterior trim has been caulked to the siding. To prevent damage to the trim or the siding, use a utility knife to score around the trim before trying to pry it off. To remove trim, use a flat-bladed prybar.

Start at one end of the trim and move progressively toward the other end, prying the trim up from the siding. It's not a good idea to start prying at the middle of the board because it takes more effort to pry and because the board can suddenly let go and slap you in the face. As the sequence of photos on the facing page shows, simply work your way around the window prying off the trim pieces as you go.

Next, move to the inside and remove the trim in the same fashion.

### SAVING THE TRIM

When removing the interior and exterior trim from a window (or door) you can make the reinstaller's life much easier if you save the trim for reuse. Because the trim fits the window exactly (assuming it was properly installed in the first place),

*Using a utility knife to score the caulking on the exterior window trim makes it easier to remove the window and minimizes the damage to the trim and siding.*

*Starting at one end,* work your way around the window, prying off each piece of trim as you go. Work in baby steps, otherwise you may break the trim.

*Interior trim* is handled in the same way as the exterior trim.

Mark the package of trim to match the window or door that it came from. A simple ID system can be used—for example, "Interior trim, Window #5," "Exterior trim, Window #5." If you are packaging trim from an interior door, mark all pieces from one side of the door with an "A" and the other side with a "B" to avoid mixing the pieces. Don't forget to mark the corresponding window or door with the number. Also, don't put your ID (even if a sticker) on the outer face of the trim. The glue on adhesive stickers can pull off original varnish finishes, and the trim piece ID in indelible Sharpie® ink doesn't look too appropriate on that varnished face when reinstalled.

## TAKING OUT THE WINDOW

After removing the trim from the interior and exterior of the window, it's time to grab the reciprocating saw. A metal-cutting blade or embedded nail blade works best. Cut your way around the window through the nails holding the window in place. There should be no need to saw into the surrounding wood. Because the blade we had in the reciprocating saw at the time was only about 4 in. long, we had to cut from both the interior and the exterior to reach all the nails. A longer blade would have eliminated this step. Once freed, the window was pushed from the inside out, with a helper on hand to ensure safe removal.

This house had been remodeled and contained a few modern double-hung windows as well as the original windows. The procedure for removing these units is the same as for the fixed-glass unit, with the exception of windows with an extended sill. The extension of the sill is part of the window frame, and for this window

*For vintage doors and windows with a nice old finish, it's important to keep the door frame and trim together for reinstallation because the original patina of the finish and the molding profiles can be difficult and expensive to duplicate.*

reinstalling the same trim saves the purchase, measuring, cutting, and finishing of new trim. This is especially important for vintage windows and doors that still have their original varnish and an aged patina, which can be difficult and expensive to match with new trim.

Denail and then stack the trim from each window and door and shrinkwrap both for protection and to keep the matching pieces together in an individual package. Bundle the interior and exterior trim separately if there is any chance of intermixing the pieces and lay the shorter trim inside the package with the longer trim on the exterior. If you have vintage trim that is easily damaged, turn the outer face to the inside of the package to prevent damage to the good face during handling and transport.

*After removing the interior* and exterior trim, use a reciprocating saw to cut the nails holding the window in place.

*You may need to cut* from both sides of the window if the blade in the reciprocating saw isn't long enough.

*After the trim* is removed and the nails are all cut, carefully wiggle the window inside the rough opening to free it, working the sash back and forth and side to side. Once it's free on all sides, have a helper on the inside push it out to you. If the window is particularly large or heavy, you might need an extra person on the outside to hold the window.

**Save the Sash Weights**

If you are removing an older double-hung window, remember that there will be two iron counterweights, called sash weights, which hang by ropes in a cavity on either side of the window. Once you remove the window trim, it's simple enough to pull out the weights for recycling or reuse.

type you want to remove only the small bottom trim piece. Bear in mind that this type of window must first be removed from the outside of the house then either rotated and brought back in through the window or brought down the ladder. (The fixed window shown in the previous example could have been taken out from the inside.)

Most of the replacement windows in the house were single-pane aluminum. Though these were not worth salvaging, each window was removed whole. The removable glass panels and screens were taken out first, and the glass panels were thrown in the roll-off (which was mostly empty at this early stage of the project) to safely smash the glass. Once broken, it was easy to

pull out the rubber weatherstripping and all clinging shards of glass. The aluminum frame, now free of glass and rubber, was then moved to the metal recycling pile and the broken glass remained in the container. Breaking the glass in the roll-off saved having to clean it up at another location.

Once the removable glass panels were taken off, we pried the window frame from the wall framing, taking care not to break the fixed panel to avoid a mess of broken glass on the job site. Then we threw the whole unit into the trash container, breaking the glass and using the same procedure as before to remove the nonaluminum materials.

One final type of window in this house was a 1970s-era greenhouse window over the sink in the kitchen. Because it was unique, the resale value was unknown. Thus we took some time to remove the window. Unfortunately, as the final nail was being pulled, the prybar slipped and cracked a glass panel, ruining any resale value. Accidents happen.

*Windows that have an extended sill (as on the double-hung shown here) must be removed from the outside of the house. Remove the small trim piece below the sill and save with the rest of the trim.*

*Because the aluminum windows* in this house had no resale value, the glass was broken out and the aluminum recycled.

*Remove the weatherstripping* before recycling the aluminum window frame.

*Use a hammer* and flat bar to pry off the fixed frame of an aluminum window and remove the whole unit for recycling.

## REMOVING A DOOR

Taking out a door is much the same as taking out a window. As with a window, our aim was to cut out the door as a unit so that it could be prehung into a new opening. As shown in the photos on this page, first remove the trim from both sides of the door and then use a reciprocating saw to cut the nails holding the door in place. Before moving the unit, be sure to close and latch the door to keep it from racking out of shape. If the sill of the door is part of the door frame (usually the case for an exterior door), you might need to cut underneath the sill to free any fasteners securing it to the floor. Also, if the door has a key lock, keep the key with the lock—if you are lucky enough to find it.

*Removing a door is very similar to removing a window. First pry off the trim, working from the end of the boards in baby steps to avoid cracking the trim. Next, grab a reciprocating saw and work your way around the perimeter of the door, cutting all the nails that secure the door to the wall framing (be sure to remove the cut-off nails). Finally, tilt the door out and remove the door and frame as a unit (typically a two-person job).*

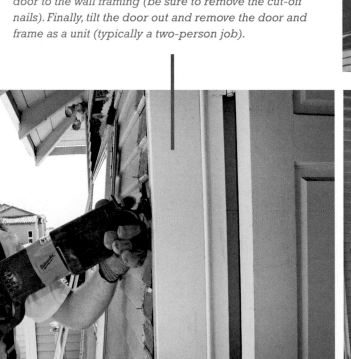

In this house, we found a relatively new prehung pocket door in one of the rooms that was worth reselling. In this case, you want to remove not only the door but also the pocket framing into which the pocket door slides. The procedure for removal is the same as for a hinged door, though here the drywall had to be removed in the pocket area to access the fasteners holding the door in place. If you are taking out a vintage pocket door, it will probably be built in. You'll have to figure out how it is hung and possibly tear out lath and plaster in the pocket area to access the rail and rollers holding the door. Be sure to save the rail and any other hardware associated with the door.

*Pocket doors* are a little more complicated to remove than hinged doors because the drywall or plaster covering the pocket has to be removed to access the fasteners holding the door frame in place.

**HERE'S A TIP**

## Reinforce the Door Framing

Although an exterior door usually has a complete frame around the door, an interior door doesn't have a sill. When you cut out an interior door, the latch side of the frame will tend to flop around and tear the frame off the door as you handle it. It's a good idea to finish-nail the frame to the door at the lower portion of the latch side to keep the frame tight to the door for handling.

Ted Reiff is cofounder of The ReUse People, a company that has diverted thousands of tons of reusable building products from the landfill into reuse.

"Our mission is to reduce the solid waste stream and change the way the built environment is renewed by salvaging building materials and distributing them for reuse."

# Floodwaters Yield a Vision of Reuse

The ReUse People of America, Inc. (TRP), a nonprofit environmental organization, was conceived in 1993 in the wake of an unusually wet West Coast winter. Heavy rains and mudslides had washed out entire neighborhoods in Tijuana, Mexico. Hundreds lost their lives and thousands were left homeless.

After the rains subsided, TRP co-founder Judy Bishop organized Project Valle Verde, a donation drive for used building materials to assist in the Tijuana rebuilding effort. Over one mid-April weekend in 1993, she and hundreds of volunteers collected more than 400 tons of new and used building materials. The aid helped build 800 houses.

While the memory of Project Valle Verde gradually waned, the vision survived. That vision—of large quantities of used building materials flowing through regional, national, and international markets—led directly to the founding of TRP, with Ted Reiff at its helm. "Our mission is to reduce the solid waste stream and change the way the built environment is renewed by salvaging building materials and distributing them for reuse," says Ted.

Ted grew up in the Midwest and saw the many farmers who never threw anything away—not a board, gate, hinge, latch, or any other building material. "Most materials outlived their original function, but the cost of acquiring new was often too great."

He came into this field with a strong business background, including experience as a retail systems consultant for an international consulting company specializing in distribution systems and as a managing partner for an investment banking firm that provided a variety of services to young technology companies. A self-described "environmentally sensitive numbers guy," Ted was drawn to the challenge of creating a viable business model within the fledgling reuse industry.

"We talk a lot about environmental sustainability," he asserts, "but my job is to create a sustainable business. Anyone can save something and keep it out of the landfill. Contractors save stuff all the time. Their garages are full, their backyards are full. Some even rent storage lockers. But eventually they throw most of the materials away because they run out of room. Few organizations are ready to absorb large quantities of used building materials. So the challenge, as I see it, is to move salvaged materials to markets where they can be reused."

Ted's long-term vision is two-fold: institutionalizing the concept of building-materials reuse and taking advantage of economies of scale. "Right now we're a cottage industry," he points out. "A few little players are trying to do something on a local basis, and they are limited by the size of their markets. Reuse is not a mainstream value. Not everyone is thinking about it—yet. And only so much can be absorbed in smaller towns. At the same time, we have to solve the distribution problem. Exactly the same challenge faced the auto parts industry 50 years ago."

In the 14 years since its founding, TRP and its affiliates have deconstructed residential, commercial, and industrial buildings ranging from single-family residences to military housing complexes and large-scale movie sets. TRP has diverted over 230,000 tons of materials from landfills in California alone.

## Salvaging architectural items

Although this house was not particularly rich in architectural features, a few items were salvaged, including the front porch columns and brackets as well as the porch railing, stair rail, and posts. If you do find nice architectural elements, take extra time and care to remove them. These are typically the most valuable individual items found in a house, and you will want to preserve their value for resale.

The first step was to cut the railing from the porch columns with a reciprocating saw, taking care not to damage the posts by cutting too close to them. Although you want to preserve as much rail length as possible, it's better to lose 1 in. of rail at either end rather than damage the post. The rail will likely have to be recut to fit a new space anyway. The handrails of the front steps were cut out in a similar way as the porch rails, except that the bottom rail was nailed down to the beam supporting the stairs. A long cut was made between the bottom rail and the beam to free the rail section.

*The original front step* railing and porch posts and brackets had architectural appeal, so extra time and care were taken to minimize damage when removing them.

*Use a reciprocating saw* to cut the front porch rail from the posts, taking care not to get too close to the posts to avoid damage.

*To free* the front porch step handrail, a reciprocating saw had to be used to cut the nails holding down the bottom rail.

*Because these posts* were set directly on the concrete of the bottom step, the bottom ends had rotted.

*Before the porch* column brackets could be removed, a utility knife was used to cut through the many layers of paint that had built up around the brackets.

Unfortunately, after we'd removed the handrail we found that the lower end that tied into the stair post was rotten. The bottom ends of the stair posts, which were set directly on concrete, were rotten too. We cut off the bad ends; though shorter than the original, the posts were still salable. We used a small trim bar to remove the decorative porch column brackets. Because they were fastened to the post with finish nails, it was relatively easy to pry them off without breakage. The porch columns were structural and supported the weight of the porch roof, so they were not removed until the porch roof was gone.

## Salvaging plumbing fixtures

A relatively new fiberglass shower unit and the attached shower fixtures were of sufficient value to remove for resale. First, the showerhead and interior fixtures were carefully removed and bagged. A fiberglass shower unit of this type has a nailing flange around the perimeter which needs to be exposed to access the screws holding the unit in place. We hammered out the drywall around the perimeter and unscrewed the unit from the walls. For easier access, we cut out a couple of studs above the unit with a reciprocating saw.

**TOOLS AND TECHNIQUES**

### Saving Plumbing Hardware

If a shower or tub is salvageable, saving all the hardware necessary to replumb the fixture may increase the resale value because the buyer will not have to purchase replacement hardware. If easily accessible, gently knock a hole in the drywall to access the shower plumbing behind or to the side of the unit. Cut the water supply lines to access the water-mixing valve and save with the other hardware. Leave a couple of inches of pipe attached to the valve(s) so the reinstaller has enough length to remove the pipe from the valve. This is especially useful if the piping is copper because the reinstaller just needs to cut the pipe off square, slip on a coupler, and then solder in place.

*A fiberglass shower unit* was saved by first knocking out drywall around the unit, unscrewing the unit from the stud wall frame, and then cutting through the front studs. Wrestling the shower unit from its home required some prying, tugging, and pulling.

Remember that in addition to disconnecting the water supply, you also need to deal with the waste drain. Usually, the polyvinyl chloride (PVC) pipe draining the shower sits a little proud of the subfloor, making it difficult to slide out the shower without first prying it up from the floor. Rocking the shower unit back and forth helped free it from the wall and the waste drain.

We were also able to save a bathroom vanity and sink, which we removed in a similar way as the kitchen sink. However, because the countertop was tiled we had to break off the tile with a hammer to remove the vanity.

*The tile backsplash and vanity top had to be broken up to allow removal of the bathroom vanity.*

## Taking up the finished flooring

The finished floors and subfloors in this house were solid Douglas fir, in good condition and well worth salvaging. It is highly likely that you'll find Douglas fir in West Coast houses as this species is indigenous to the region and is common in homes built before the advent of plywood (which started to be widely used after World War II). In the Midwest and Eastern United States, hardwoods such as oak and maple are more common.

Except for the oldest homes in this country, where you can find solid-plank flooring, most homes have tongue-and-groove (T&G) flooring. The advantage of a T&G profile is that the interlocking joint keeps the individual planks flat while allowing the wood to shrink and swell with seasonal changes in humidity. The T&G configuration also requires a specific removal procedure to minimize damage to the flooring.

**When removing flooring:**

1. Always start from the tongue side.

2. Pry close to the nail.

3. Try to pry up and out to minimize damage.

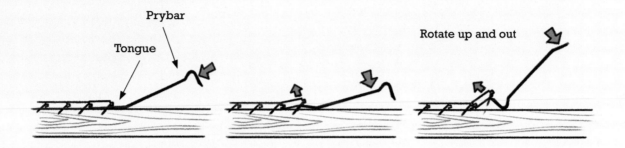

Prybar

Tongue

Rotate up and out

*The Douglas fir* finished flooring in this house was in good condition and worth salvaging for resale.

*The profile* of tongue-and-groove wood flooring allows the individual pieces to interlock, which helps keep the floor flat. The flooring has to be removed in the reverse direction of the installation.

*You'll probably sacrifice the first row or two of flooring to gain enough room for prying.*

For all tongue-and-groove flooring, the basic procedure is as follows:

**1** Determine which side of the room the tongue of the flooring points toward. If you can't see the ends of the flooring pieces, you may need to pry up a piece at either side of the room to determine the orientation.

**2** Start on that wall and remove the first piece of flooring; you should be working from the tongue side. You may have to sacrifice one or two rows of flooring to create some room to work with a prybar. Once enough room is created, you will want to work row by row toward the opposite side of the room.

**3** Start at the end of a board and gently pry the tongue side of the flooring up a bit. Work your way down the piece, prying as you go. Don't try to pry the board out in one go, unless the nails are quite loose. The key is to pry up and out, not straight upward, otherwise you will break off the bottom tab of the groove side. Try to pry as close to each nail as possible to minimize the risk of the tongue splitting away from the nail. If you encounter a really stubborn nail, try rocking the pry tool back and forth to loosen the grip of the nail. If that fails, you might have to get out the reciprocating saw to cut the nail.

*Starting on the tongue side,* pry your way down the board a little at a time. Work carefully and you won't break the tabs on the groove.

**Removing Flooring**
To remove solid-wood tongue-and-groove flooring, always work from the tongue side and pry up and out. This will prevent splitting of the flooring and maximize the yield of reusable material.

Several tools can be used to pry flooring. The flat prybar works well, especially if you have two or more people to work on longer pieces. But because the flat prybar is a short tool, you must work bent over or on your knees, which can get tiresome in a hurry. A taller prybar allows you to stand and pry the flooring and can provide more leverage.

Some unbuilders fashion their own flooring bars with a flat blade and a 90-degree bend in the bar (see the photo on the facing page). Also, some of the specialty tools described in chapter 5 work well and may be a worthwhile purchase if you remove a lot of flooring.

*Two people* can speed up the work and save on shuffling back and forth when dealing with longer lengths of flooring.

*A taller flat-bladed bar* works well for removing flooring while standing up, which is a lot easier on the knees and back.

*This home-made flooring bar is constructed with a flat blade and a 90-degree bend. A round steel bar welded to the back stabilizes the tool.*

The wrecker's adze can be used to lift three or four rows of flooring at a time. The side-to-side motion, a little at a time, levers up the flooring.

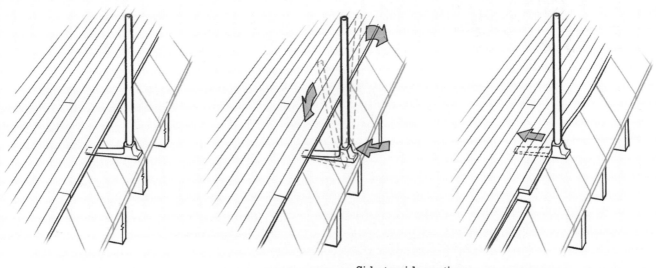

Side-to-side motion
levers up flooring.

The light fixtures, doors, windows, cabinets, and plumbing fixtures removed during the soft strip were stockpiled in the living room of the house for convenience of loading and transport. A truck was scheduled to arrive when the soft-stripping was finished to transport these items to the reuse warehouse. (The flooring went later with the rest of the items salvaged from the full deconstruction.) Fortunately, all the items removed fit in one load. We used carpet and carpet padding from the house to cushion several items; but even with padding, proper loading of the truck is paramount to prevent breakage.

Never lay windows flat. Stack three to four deep in the same orientation as they were in the wall and lean then against the outside walls of the truck. Doors can be stacked over the windows for protection. Base cabinets can be loaded down the middle of the truck, with padding placed between the units and the doors and windows. Stack the upper units on top of the lowers with padding between.

*At the end* of the soft-strip, the salvaged items were loaded into a moving van for transportation to the reuse warehouse.

## Cleaning Up

While we all know that a clean job site is a safer job site, in the frenzy of activity messes are inevitably made. Although we did a general job-site cleanup at the end of each day to minimize safety hazards, we also stopped periodically to do a more thorough cleanup. This is conveniently done at logical stages of the deconstruction process, such as at the end of soft-stripping or after the interior finishing material has been removed. Cleanup provides a psychological boost to everyone on the job because working in a clean, organized space is always more pleasant than working in a cluttered, messy area.

And there you have it, a soft-stripped house. Some people would stop at this point, satisfied with salvaging the easy-to-get items. If you are more adventurous, read on. Next, we tackle the second, and more involved, phase of this project—deconstructing the rest of the house, down to ground.

*It's important to stay on top of cleanup and debris removal to keep the job site safe and more pleasant to work in.*

## Chapter 7

# WHOLE-HOUSE DECONSTRUCTION

*Bite off more than you can chew,
then chew it. Plan more than you can do,
then do it.*
—Anonymous

In the previous chapter, we covered soft-stripping the house of the materials that are relatively easy to salvage, including cabinets, lighting, doors, windows, and finished wood flooring—the icing on the cake. The next level of harvest, taking the whole house down to the ground, is more involved. However, the rewards can be great, as most houses will yield thousands of board feet of framing lumber, sheathing, and other building materials. A full deconstruction is a big job, and even with lots of planning, it will still be a big job. The only way to eat the elephant is to start eating... so, let's do it!

Before you can harvest the lumber in a house, the walls, ceilings, floors, and roof coverings must be removed. Basically, the process of a full deconstruction continues the last-on, first-off sequence, taking off the interior wall and ceiling materials, the siding, and the roof shingles to expose the sheathing and framing of the structure for disassembly and removal. After everything is removed, the final steps are denailing, trimming, and stacking the wood.

*A full deconstruction continues the process begun with the soft-stripping, taking the whole house down to the ground. Even a small house can yield several thousand board feet of framing lumber and other material.*

*The sequence* of a whole-house deconstruction can vary, but it usually follows the reverse order of construction, starting with removal of the interior finishes, non-load-bearing interior walls, and the roof, and then exterior siding, exterior wall and floor framing, and the foundation.

In this chapter we continue the house deconstruction started in the previous chapter and go through the steps we used to take this structure to the ground. Every house is different and not every tool technique or material removal option can be illustrated in the deconstruction of a single house, so, along the way, we will interject tool tips, alternative methods, and safety pointers that should help you in other situations.

Don't get too fixated on the sequence we follow in this house. Because this house was deconstructed during the summer in California, the chance of rain was remote. For that reason, we chose to remove the roofing materials first, then the roof rafters and second-story shed dormer, followed by the first-floor ceiling joists, wall coverings, first-floor siding, and first-floor framing. The rationale was that by removing the roof first and opening up the top of the house, we'd get more natural light and ventilation into the building.

Obviously, in other parts of the country where rain is more frequent (or in California in the winter), a different approach might be taken. Where rain is likely, it makes sense to remove the roof as late in the game as possible to keep working conditions and the interior building materials dry. In wetter climates, we would probably remove the interior and exterior wall materials before taking off the roof. The point is that you have some flexibility in the order in which you take the building apart, weather and available workforce notwithstanding.

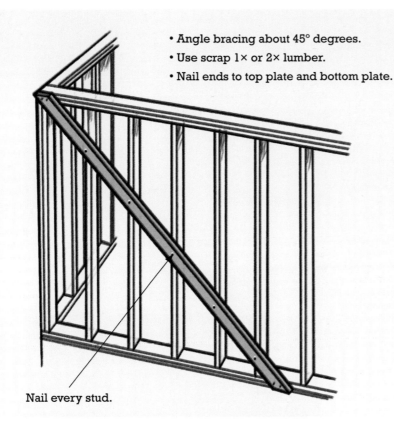

**TEMPORARY WALL BRACING**

- Angle bracing about 45° degrees.
- Use scrap 1× or 2× lumber.
- Nail ends to top plate and bottom plate.

Nail every stud.

**HEADS UP**

**Leaving the Roof until Last**

If you decide to save the roof until last in your deconstruction sequence, remember that leaving a heavy roof on a spindly building frame with no interior or exterior sheathing can create an unstable structure. We suggest putting up some temporary bracing on the exterior walls in the form of long 1×s or 2×s nailed to the wall studs at an angle across the wall. This should be done as the exterior sheathing is removed from each wall. The bracing should extend from the top plate to the bottom plate. On a long wall it is better to have a few shorter braces rather than one long one.

## Maintaining the Building's Integrity

To state the obvious, the most important consideration in doing a full deconstruction is that you keep the building from falling down on top of you. The quickest way to a disaster is (a) to remove load-bearing elements (walls, beams, posts, and so on) while they are carrying load or (b) to remove some element that maintains stability in the building frame, such as shear bracing. In most conventionally framed buildings, if you progressively remove materials from the top down (roof down to foundation), you will remove the weight carried by a load-bearing element before you remove the element itself, minimizing the danger of collapse. There are exceptions to this, such as when the building is built with overhangs, cantilevered beams, or offset walls, which can become unstable if you remove stabilizing counterweight or structural connections. A knowledgeable builder or engineer should be consulted if your building is framed in an unconventional manner.

Although it is ideal to work from the top down, there may be times when you'll want to take out an interior wall or part of an exterior wall to provide improved access or make it easier to move materials through the building (for example, cutting into a gable-end window wall, as shown on pp. 164–165). In this case, you have to determine if the wall in question is load bearing or not. Typically, if an interior wall runs parallel to the joists above it, it is not load bearing. If it runs perpendicular, it *may* be load bearing, depending on the span of the joists above (see "Know Your Bearing Walls" on p. 90). There are exceptions, so don't start cutting out studs until you know for sure if what you are cutting out is carrying load or stabilizing the structure. If in doubt, call in a builder or engineer.

## Panelized Deconstruction

It doesn't make sense to take apart the whole building if you're going to put it right back together again. In some cases, you can disassemble a building into panels and reassemble it on another site. That was the case with the garage shown here. Using hand labor, two people removed the roof and each wall of the garage and erected it on a new site. They did have to add one wall and used siding salvaged from the main house (which was being torn down) to match.

*This garage* was taken down in sections and reassembled on another site.

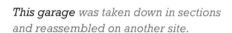

## Roof Tearoff

Roofing is rarely worth saving, unless it is tile, slate, or corrugated metal. Even newer wood shingles are difficult to remove for reuse because they tend to split when pried off. The first task before sending a crew up on the roof is to inspect it to make sure it is sound enough to walk on, especially if roof shingles are missing or you've seen evidence of leaking in the attic.

On this house, we installed a safety cable across the roof to serve as a tie-down for individual safety harnesses. This involved screwing down a ringed bracket directly into the roof rafters at each end of the highest points of the roof (at the top of the shed dormer) along the ridgeline. Attaching individual safety lines to this cable and to personal safety harnesses made work on the roof secure and comfortable. We used a spotter to help manage the safety lines and ferry tools to the workers when several people were on the roof.

Relatively new rolled asphalt roofing had been installed on the shed dormer. We used a utility knife to cut the roofing into sections about 5 ft. wide, pried up an edge with a roofing shovel, and then pulled up the sections by hand. The manageable pieces of rolled roofing could then be rolled off the roof or pitched into the roll-off trash container below.

*Metal brackets* were installed at each end of the shed dormer roof, and a stout steel cable was attached end to end.

*Each worker* was tethered by safety rope to the steel cable, allowing access to the entire roof.

*A co-worker* served as a spotter at the cable to shorten or lengthen each worker's safety rope as needed.

Aside from the dormer, the rest of the roof had a layer of wood shingles overlaid with two layers of asphalt shingles. The top layer was near the end of its life and very brittle after years baking in the intense heat of the California sun. Normally, we would recommend the use of a roofing shovel to remove asphalt shingles,

## Stripping Shingles

When removing shingles, let gravity do as much work for you as possible. Work from the top down and try to funnel materials down the valleys of the roof directly into a roll-off trash container or truck. Roofers who do complete tearoffs use temporary scaffolding, plywood chutes, and other job-site-built ramps to maximize the use of gravity to get the old roofing to a roll-off or truck. Pushing roofing debris directly off the roof onto the ground only means you will have to pick it all back up (working now against gravity) to load it into a container.

*Gravity* is your friend. Use it when you can.

*A plywood ramp* or chute can help "shoot" debris into a roll-off that you can't get right next to the building.

A pickax worked well on this roof. By inserting the pick end between the board sheathing, workers were able to pop up the brittle wood shingles.

## Using a Roofing Shovel

Roofing shovels are designed to easily remove asphalt shingles. They are available in a variety of styles, though each has a flat blade with teeth and either a heel on the blade or an angled handle that allows you to lever the blade once it has hooked under the shingles.

The key to working this tool is to use the teeth to pull the roofing nails out along with the shingle. This will make de-nailing the roof sheathing a lot easier and will help eliminate the nail heads, which snag your pants, sleeves, boots, and various body parts.

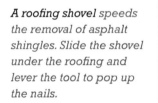

A roofing shovel speeds the removal of asphalt shingles. Slide the shovel under the roofing and lever the tool to pop up the nails.

but on this roof a pickax and framing hammer worked well due to the brittleness of the roofing. Because the original wood shingles were laid over spaced board sheathing (a normal practice for wood shingles), it was easy to slip the pick under the shingles between the sheathing boards and pop up each row of shingles.

Asphalt shingles can be slippery especially as they get old and brittle and they begin to break down, so this work is not for the unsure of foot. Though you should be extra cautious on any roof, it gets much more difficult to walk on roofs that have a pitch greater than about 5/12. A 5/12 roof slope means that for every

*Using a roof jack—a metal bracket sized to fit a 2×6 and nailed to the roof—makes it possible to work on steep roofs. Make sure to nail the jack directly into a rafter.*

12 in. along the horizontal (run), the roof rises 5 in. along the vertical (rise). If you feel uncomfortable with the steepness of the roof you're working on, we recommend nailing 2×4s parallel to the ridge to provide extra footholds or using roof jacks. These footholds are also helpful to use as places to set larger hand tools and as a stop in case you drop one and it starts to slide off the roof. Of course, you should be wearing your safety harness.

## Getting onto the roof

Depending on the height of the house, the ground around the house, and access to the attic, there are typically three ways to get onto the roof. The first is on a ladder directly from the ground. This can be very convenient for a one-story roof but is much less so for anything higher.

The second option is to access the roof from second-story balconies or porches or from the low-sloped roof of a first-story porch. This can greatly shorten the distance you have to climb on a single ladder. You might be accessing the porch roof from a window (first cutting out the sill) or from a ladder from the ground. In the overall planning of the sequence of deconstruction you may find a porch roof a good place to access upper-story windows and siding, so use these work platforms to your advantage for the higher levels of the house.

The third way to access a roof is from the attic, by cutting a hole in the roof. If the house has living space under the roof accessed by a stair, you may be able to avoid ladders altogether and have a convenient place to stage tools within easy reach.

After we removed the roof shingles on this house, we moved inside to attack the walls. The house had been remodeled over the years and contained a mixture of interior wall finishes. The added dormer was gypsum drywall, and the first floor was lath and plaster. If you use the right techniques, removal of both types of finishes is relatively easy.

### Lath and plaster removal

Lath and plaster is an old type of interior finish that's found in most homes built before World War II. The most common type of lath is wood, though metal lath is also found (and is more difficult to remove). Although lead-based paint is more likely on exterior wood siding than on interior plaster, it's a good idea to test for lead because removing plaster creates a lot of dust. Chapter 5 explains how to use a lead test kit.

The common mistake in removing wood lath and plaster is to use a claw hammer or sledgehammer and try to remove both materials in one step. This method is slow and exhausting, and the head of a hammer is simply

too small for the job; all you will do is punch holes in the plaster and break the wood lath. Also, you end up with a pile of intermixed lath and plaster that is difficult to shovel and bulky to move. A much more effective method is to remove the plaster completely, clean it out of the room, and then come back in and remove the lath. Pulling off the lath in a separate step allows you to more compactly bundle the lath for efficient removal and possible recycling.

*Plaster removal is dusty. Make sure to wear a disposable respirator!*

*By removing plaster and lath in two steps and keeping them separate, it's easier to move both materials out of the house for disposal.*

Start the plaster (and drywall) removal process by creating maximum ventilation—such as opening or removing windows and doors, if they haven't already been removed. A flat-bladed or roofer's shovel works well to remove the plaster. Basically you want to hit the wall to crack the keys off the back of the plaster then use the blade of the shovel to remove sections of the plaster face. The idea is to make the most of each hit by aiming each blow to a spot midway between the underlying studs. This flexes the plaster and more effectively cracks it. It's important to hit the plaster hard enough to crack it, but not hard enough to crack the wood lath, which will only make the lath harder to remove later. It helps to shovel off as much as possible in a direction parallel to the lath strips, so you don't constantly catch the tip of the shovel in the gaps between the lath.

Place a wheelbarrow against the wall directly below the plaster to eliminate at least some of the work of scooping it off the floor. Alternatively, line part of the wall with several

*Hitting the wall with the back of a shovel will crack the plaster. Work up and down between the studs.*

*With the plaster cracked, it's easy to shovel off big chunks. Work parallel to the lath and the shovel tip will catch less often.*

wheeled rectangular plastic trash cans, which can be rolled outside to dump (be careful not to fill them so full that you can't move them—plaster is heavy!). Work around the room and finish the walls before tackling the ceiling.

Waiting to do the ceiling till last means you don't have to contend with dropped plaster all over the room while you're working on the walls, minimizing a slip and trip hazard on the floor and making it easier to move the wheelbarrow. It's harder to get the plaster off the ceiling than off the walls as now you have to swing a heavy shovel above your head. Standing on a rolling scaffold makes the job easier because if you are closer to the ceiling you can more effectively hit with the flat of the shovel, and it is easier on your shoulders and back. If you have access to the ceiling lath from the top side—that is from the attic or second floor above—you can make the plaster removal job easier by whacking the lath between the ceiling joists in the same manner (except now from the backside) as described for the walls. Because you have gravity and the weight of the plaster working for you, much of it will fall from the ceiling, eliminating the need to use a shovel to remove it from the ceiling below. A flat-headed shovel and a scoop shovel work best for cleaning up the plaster.

Once you have the plaster removed and cleaned up, it's time to start with the lath. Ideally, you want to remove each piece of lath along with all its nails. Lath is held to the wall with small nails, so you can simply hook a tool behind the lath and yank it from the wall. If you hook your tool close to the stud, you will break less lath. Because the lath is pretty flexible,

*A flat-bladed shovel* works well for cleanup. *If you can handle the weight, a scoop shovel carries more debris.*

you have to develop a feel for pulling hard enough to pop the nails out, but not so hard you break the lath.

There are a couple of tools that work for removing lath. A flat-bladed prybar is small and easy to handle and has about the right length of blade to hook behind one lath at a time. It also allows you to pull nails as you go. Some people prefer a bigger tool, such as a short-handled pick or mattock or a long-handled shovel. Though heavier, a larger tool allows you to pry against several lath strips. Crowbars don't work well for lath removal as the curve of the bar is too tight and all you'll do is break the wood.

*Remove the lath* and drop it into a wheelbarrow when possible. Try to pry the lath gingerly enough to pull as many nails out as you can with the piece of lath. If you pull too hard, the lath will break at a nail. Then not only do you end up with a lot of small pieces of lath to deal with but you have to go back and pull a bunch of lath nails from the framing.

*Once you open up* part of the wall, another option is to use a longer tool, such as the shovel shown here.

Once all the lath is removed, pull out any remaining nails. We prefer to use a flat prybar rather than a hammer for this job as the small lath nails tend to get stuck between the claws of the hammer, requiring you to stop to clear each nail.

## Drywall removal

If you're dealing with drywall, try to remove it in the biggest pieces easily handled by one person. Drywall is somewhat frustrating to remove, as it always seems to break where you don't want it to. To get started, use a flat-bladed shovel, garden spade, or roofer's shovel to make a horizontal cut at about shoulder height. This helps create a place you can grab the drywall and pull it down in pieces. For the upper part of the wall, place

### Pulling Lath Nails

Pull all the nails you can after removing the lath. It's easier to do it while the stud is upright and held in place rather than at the denailing station, where you will have to wrestle with the stud with one hand while trying to use your prybar with the other. If you break the head off a nail, use a pair of end-cutting pliers or nippers to grab the remaining nail shank.

*When removing drywall,* try to pull off pieces easily handled by one person.

## The H Method

Recycle North, a deconstruction and building materials reuse organization in Burlington, Vermont, removes drywall using what they call the "H method." Here's how it works:

Start at each end of the wall and knock out a vertical line from ceiling to floor about 6 in. in from the corner. Then knock out a horizontal line about 4 ft. up from the floor from one corner to the other. Grab the bottom half at the 4-ft.-high seam at one end and pull a little. Move down and pull a little bit more and repeat across the room. This can loosen an entire sheet if it was laid lengthwise (or half-sheets if laid vertically).

Repeat the process by pulling out on the upper half of the wall. Come back to the corners and knock out the vertical strip that remains.

Take off the metal corner bead on an outside corner by hitting it right at the corner with a hammer, from top to bottom. This flattens the middle of the angle and forces the edges out so they pop off of the tape and joint compound. This makes it easier to grab and pry the entire length by hand, prybar, or hammer claw.

*Once the drywall on one side of an interior wall is removed, you can use the flat of a shovel to loosen the drywall on the other side of the wall.*

the head of the spade or shovel along the wall behind the drywall, angle the handle across the nearest stud, and then pry the piece off.

Once you have removed the drywall from one side of the wall, it is easy to remove it from the other side of the studs if you push it off from the backside. Some people use a garden rake to pull drywall off by hooking the tines of the rake behind the edge of the piece. In any case, avoid using tools with a small head, such as a claw hammer or a sledgehammer. You'll only create a mess and wear yourself out.

## Removing Electrical, Plumbing, and Ductwork

With the interior walls and ceilings removed and at least the interior skeleton of the house exposed, it's a good time to start removing accessible electrical wiring, water pipe, waste lines, ductwork, and other items that will get in your way later as you start to take apart the building

*If possible, remove all electrical wiring and other easily accessed hardware while the framing is still in place.*

frame. Like the lath nails we discussed earlier, it's often easiest to remove these items while the framing is in place rather than at the denailing station. That being said, you don't want to risk your life crawling out on a spindly rafter to remove a piece of wire. Use common sense and remove what you can easily reach.

To remove most house wiring (12- or 14-gauge Romex), a good set of wire cutters or snips will work. Some are made with an angled head, which makes working in a tight place a lot easier. Cutting heavier wiring may require specialty cutters, available only at an electrical supply house. If you are able to recycle electrical wire where you live, you might want to have a barrel handy to throw the wire pieces into.

To remove metal plumbing pipe, get out your reciprocating saw and cut the pipe into manageable lengths for transfer to the recycling pile (cut at the threads and the cutting will go faster). If you do an industrial deconstruction and encounter lots of pipe, you might want to use a gas-powered circular cutoff saw with a metal blade. Don't waste your time trying to unscrew old pipe—it's going to be recycled anyway. PVC pipe is very easy to cut with a wood reciprocating sawblade, and cast-iron waste stacks can be broken into manageable pieces with a sledgehammer.

Be wary of water remaining in pipes and waste traps. If you cut into a full waste trap while working over your head, no one will want to sit next to you at lunchtime. Also, as you remove traps, you will allow sewer gas to escape from the sewer main. If you are doing a whole-house deconstruction, you might want to go into the basement or crawl space before work begins and cut the main waste stack at that point and cap the opening to keep the fumes from migrating through the house for the rest of the project. If you're doing soft-stripping only, plug any waste lines you open at the fixtures to keep the smells contained.

## Removing Roof Sheathing

The next step on this house was to go back on the roof to remove the 1×4 board sheathing. Unlike construction today, where oriented strand board (OSB) or plywood sheathing predominate, it is very common to find spaced board sheathing in prewar homes. The spacing was there for a reason: to allow the wood shingles to dry out from the attic side of the roof. The 1×4 board sheathing on this house was up to 24 ft. long, a testament to the type of materials available in 1901.

Plywood or OSB sheathing is common on newer buildings and can be difficult, but not impossible, to remove. A sheet usually has a lot of nails in it (and in floors it is often glued down). If you have access from below, try knocking the sheet up near the rafters with a sledgehammer to pop up the nails, then have a co-worker tap the sheet back down, leaving the nail heads still protruding. The idea is to get the nails popped up enough to remove them with a prybar.

*Very long pieces* of board sheathing were recovered from the roof.

*Removing the sheathing* allows access to the roof rafters.

*Spaced 1×4 roof* sheathing is easy to pry off using a wrecking bar.

## Removing Shingles and Sheathing as One

TOOLS AND TECHNIQUES

In some cases, the board sheathing under the roof shingles isn't worth saving. If you can determine this before removing the shingles, you can save some labor by removing the shingles and sheathing in one step. The idea is that you work across the roof and pry up and roll over the sheathing and shingles together, letting gravity pull the combined bundle down the roof slope and into a roll-off container below.

The first step is to remove the ridge-cap shingles. Next, use a reciprocating saw to cut through the roofing and sheathing midway between rafters in a vertical line between the ridge and the eaves. Starting from the ridge you can pry off both the sheathing and roofing in one step. (The Board Lifter shown on p. 129 works well to lift the sheathing off the rafters.) The bundle of material you have to roll over is as wide as the spacing between each cut, so space the cuts according to the number of people you have to help.

When possible, work from inside the roof by laying plywood over the ceiling joists (assuming the roof slope is low enough) or use a rolling scaffold from below, if you have a floor in the attic space.

## Trimming Lumber in Place

It's often more efficient to trim a piece of lumber in place. Where several pieces of lumber join, such as at the end of the hipped gable roof in this house, there can be many nails and angle cuts. Rather than try to pry each piece apart, pull all the nails, and trim the piece square, it can save time and effort to simply use a reciprocating saw, circular saw, or small chainsaw to cut the rafter as close to the joint as possible. You will sacrifice very little lumber and you will save much labor removing nails (often in ends of a board that will be trimmed anyway) at the denailing station.

*The damaged* and nail-filled end of this rafter was trimmed in place, saving a step in denailing later.

## Removing Rafters

The 2×4 rafters were the first structural lumber to be removed from this house. Because the roof was a hipped gable, the ends of the house produced many shorter pieces of lumber. All 2×4s longer than 6 ft. were saved for resale, while a pile of pieces between 4 ft. and 6 ft. was saved to use as bolsters in stacking lumber. The rafters were pried apart using a prybar at one end. Where the rafters were long enough, the other end could be pried apart using the rafter itself as a lever. Be careful at this point because you can split and crack the end of the lumber by being too forceful. Being gentle doesn't take but a few seconds longer, and you will end up with less damage at the end of the lumber. Where several pieces of lumber came together in this hip roof, we used a saw to cut rafters close to the joint. This saved time denailing later. In the end, the length of lumber lost was negligible.

Rafters in older buildings are typically attached to the top plate by toenailing. Newer buildings may have metal straps, often called wind (or hurricane) clips, nailed into the roof truss and to the top plate or wall studs. If the rafters have a bird's mouth cut over the bearing wall, it may be easier to cut the rafter just inside the bearing wall. You lose some length of lumber (the exposed rafter end), but because you'll probably have to trim out the notch later to resell the lumber, it's more efficient to trim it on the roof.

*Because this roof* was a hipped gable, the length of the rafters varied.

*Freeing one end* of the rafter with a prybar allows you to use the lumber as a lever to remove the other end. Be a little gentle at this point, because you can easily split and crack the end of a brittle rafter if you use too much force.

*In older homes,* the roof rafter was often cut with a bird's mouth to bear on the wall top plate. This notch has to be cut out for the lumber to be reused, and it's usually easier to do it while you are on the roof.

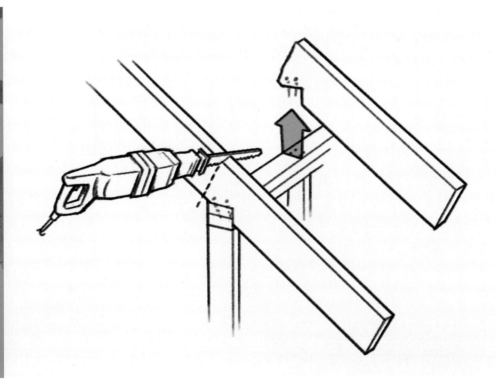

*Jim Stowell of the Whole Log Company started sawing timber to build a house and turned it into a career remilling heart pine flooring.*

"Cutting into those timbers is like diving
for buried treasure and coming back up
with your pockets full of gold."

# Gold from Old Timbers

When Jim Stowell bought a homestead in Zirconia, North Carolina, in 1978 his first tasks were to build a home and clear some land to plant blueberries as an income source. "I bought a small sawmill and figured I could cut up the dead oak logs scattered over the acreage where I wanted to plant and produce enough wood to build a timber-frame house. I built a nice home from that oak (and picked a lot of blueberries!), but little did I know that 25 years later I would be sawing salvaged timbers full time."

Cutting his teeth on the woods found on his land, it wasn't long before he stumbled on sources for reclaimed wood to remill. "I had the opportunity to buy an old wooden flume used in the old logging days to float logs. It was a mile long and built of cypress. It was a lot of wood to swallow at once, but we remilled it all and built countless doors and walls of beautiful paneling with it."

Living in the South, Jim soon got the chance to work with antique heart pine. "Heart pine was superior to any wood I'd worked with before . . . . strong, rich, straight, and beautiful. I was smitten."

Not that learning to work with heart pine was easy. Some of the wood was salvaged from factories over 100 years old, and it was often very dirty, with accumulations of paint, oil, and nails—often lots and lots of nails. And it ate up sawblades and planer knives like crazy. "It's hard work, but cutting into those timbers is like diving for buried treasure and coming back up with your pockets full of gold," says Jim.

Today Jim works mainly with salvaged heart pine timbers to produce flooring, stair parts, and other custom building materials in his sawmill. Flooring is the biggest market, and Jim has seen trends change over time. "While the trend in upscale hardwood flooring has traditionally focused on finely sanded, gloss-finished floors, today the distressed and rustic look in antique wood floors is increasingly popular. In either case, the beauty and richness of antique wood comes though. Our feeling is that the wood should live another life: Once is not enough."

It seems a simple matter to move the lumber you salvage to the ground: Just pitch it off the roof into a pile. This is not only unsafe (if you catch a nail on your glove as you are throwing a piece, you may follow the board to the ground), but also foolhardy because you will damage a lot of the pieces, reducing the value of the material you've worked so hard to salvage. That being said, if you are in a situation where you *must* drop lumber to the ground, you will minimize breakage if you drop the piece straight down so it lands on one end and then falls over.

When possible, it's best to lower the lumber over the edge of the building and lean it against the wall for a co-worker to move to a pile. If you are too high up, then you might lower them down inside the building for someone to ease out of a lower-story window. A high-reach forklift also works well, as the lumber can be loaded directly onto the forks for lowering to the ground.

In addition to moving lumber materials you will constantly be moving debris to the roll-off. You can lighten the load on your back substantially if you take the time to build ramps either to carry material to the container or to accommodate a wheelbarrow.

*Lower the lumber* off the building with care to avoid damage.

*Build ramps* to the roll-off container, when practical, to minimize the work of moving debris.

## Taking down Trusses

In the roof of this house, we took apart the roof framing piece by piece. In a roof built with trusses, there is the option of removing the trusses whole, either to be reused again as trusses or for disassembly on the ground.

Reusing a truss is more involved and requires large trucks to haul the long unit. Complicating this is the fact that the trusses were designed into the particular building you are taking them from, and it may be difficult to integrate them into a new use. More important, the trusses you salvage may not meet the engineering design codes for a new application because engineering design criteria have changed over the years. So you may go through a lot of trouble to salvage a truss and find out that it won't work according to new building codes. Unless you find a buyer who can use that particular truss and get approval

to use it, you may be better off disassembling it and salvaging the lumber pieces for reuse.

Depending on its size, it may be safer and easier to remove a truss whole for disassembly on the ground. If the truss is not too large, you can use a block-and-tackle or other rope-and-pulley system to lower each truss (assuming you have the room below to swing the truss). This is the reverse of the process a homebuilder uses to roll a truss up into place on a house (when a crane isn't used).

Another alternative is to cut the building in sections and move them to the ground for disassembly. This requires machinery for lifting heavy sections and probably makes more sense for commercial, military, or other open-spaced buildings than for a single-family home.

*The roof* of this military warehouse was built with rafters and a long lower horizontal collar tie, essentially making it a simple series of trusses. The deconstructors cleverly took advantage of this fact: After pulling the toe-nailing at the heel of the truss and cutting each truss away from the 2× ridge beam, they gently lowered the truss with a block and tackle for disassembly on the ground. The block and tackle was tied to the remaining roof structure.

*A high-reach forklift* was used to remove sections of an army barracks to be dismantled on the ground.

## Removing a Dormer

The dormer on this house had been added only a few years earlier, and because it was newer construction, the roof was constructed using plywood. The fascia and other trim were removed first followed by the roof plywood. This exposed the roof rafters for removal. It helped to pop up the sheets of plywood with the head of a sledgehammer from the room below to loosen the nails. They could then be removed from the roof deck above using prybars and hammers. The roof rafters of the dormer were pried apart in exactly the same way as the roof rafters on the rest of the house.

*First, the fascia* and other trim were removed from the dormer; then the plywood on the low-pitch roof was pried up to expose the roof rafters.

*The dormer rafters* were pounded and pried apart in the same way as the rafters on the main roof.

## Removing Ceiling Joists

Once the roof rafters and dormer were removed it was time to salvage the first-floor ceiling joists. Before that could be done, we shoveled out the vermiculite insulation from between the joists. Rather than try to walk on the joists, it's much safer to lay down some plywood and use that as a temporary work surface. Basically, the process of removing the ceiling joists is to cut or pry out the nails holding the joists to the outside bearing walls and any interior walls. The joists were then removed full length.

*With the roof* rafters and dormer removed, the next step was to tackle the ceiling joists. But first, the ceiling insulation had to be shoveled out from between the joists. Plywood sheets were used as a temporary walking surface.

*Each 2×6 ceiling joist* was pried from the walls they were bearing on. They were then slid over to the edge of the outside wall or through a window and leaned against the building for transfer to the denailing station.

### Removing Joists with a Chainsaw

Barry Stup, long time salvager of barns and industrial buildings, has a great tip for quickly removing joists using a small chainsaw.

At one end of the joist, cut a small notch about 1 in. deep in the *bottom* side as close to the end of the board as possible but far enough from the end that you miss nails or other hardware. Next, cut down from the top of the joist in line with the notch and leave about 1 in. of wood

remaining, as shown in the drawing below. This is enough material to hold the end of the joist in place (but don't walk on this board!).

At the other end of the joist, you want to cut at a slight angle (about 10 degrees from vertical is enough) all the way through and let it drop. Cutting at the slight angle shown in the drawing will provide room for the joist to fall away. As it starts to swing down, there is enough weight to pop the bit of wood you left at the other end and the joist will fall down.

Note that you have to angle the cut the correctly. If you make the angle cut opposite the direction shown in the drawing, the joist will hang up.

Barry says that though you lose a little of the joist at each end, this method results in less splitting of the boards than would occur if they were pried apart. It's fastest to cut all the joists at one end first, then move to the other end and drop them one after another. This method works well only with a chainsaw because you need the wide saw kerf to provide clearance for the joist to drop free.

*Note:* Make sure the floor is clean of debris. A sure way to increase board breakage is to have it land on something midlength, which will flex the board and crack it in the middle.

First cut

Second cut

1 in.

Third cut

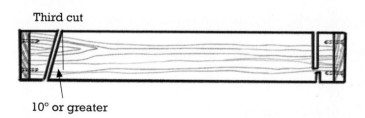

10° or greater

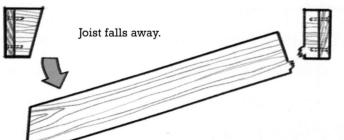

Joist falls away.

Tab of wood breaks when joist swings downward.

The siding you find on a house can vary tremendously in quality, condition, value, and difficulty to remove. The horizontal lap siding on this house was high-quality clear redwood, nominally 1×8 in size with a double-drop profile. Two options exist for this siding: The first option is to remove the siding and reuse it as is as replacement siding for buildings with similar profiles. The second option is to remove the siding with the intention of remilling the profile or flipping it over to be used again as siding (this usually works only for bevel or flat siding).

The basic difference between these two approaches is that if the siding is to be used as a replacement siding, you want to minimize damage to the existing front face of the siding. If it is to be remilled, you want to minimize damage to whatever face will be the new face. If flipped over, you would want to keep the backside undamaged. The deconstruction manager concluded that the best market for the siding on this house would be to sell it as replacement siding because the profile is common in older California homes and is unavailable unless custom milled.

*The siding on this house* was clear redwood and well worth saving. The profile was such that it was not practical to flip over and reuse as unpainted siding. (Flipping over siding for reuse works best with bevel siding.)

How the siding was installed affects how you remove it. If the siding is over sheathing, then you will have to pry it from the outside using a prybar. The procedure for removal in this case is very similar to that for removing flooring. Start at one end of the uppermost piece and gently pry the siding from the wall. Once you have it started, you can use the gap at the top to insert the prybar farther down the length of the siding, prying a little at a time to loosen the nails as you go. If necessary, work back over the length of the piece a second or third time to take it all the way off. If you get one end loose and you have some help, one person can flex the siding and the other can use the prybar to progressively loosen each set of nails. Damage to siding is usually done by trying to pry it off in one go or prying too far from the nail and flexing the board too much.

When the siding is the only material nailed to the outside of the exterior wall studs (which was the case in this house), you have the option to work from the inside of the wall, which can be very helpful, as you can do much of the work on siding high up on the wall from inside the building.

Working from the inside also offers you the access to bump out the siding from the studs using a piece of 2×4. Tapping out the siding helps you get the nails popped out enough that you can start prying. This will also give you room to use a reciprocating saw to cut off the siding nails from the

## LAP SIDING

Horizontal lap siding is laid up from the bottom of the wall to the top with each piece above lapping over the piece below. The pieces are generally nailed with two nails at each stud, the first nail through the exposed face of the siding and the second nail at the top, often underneath the lap of the piece above. You have to pry siding off with care or it will split, especially at the ends.

*With help,* you can have one person flex the siding while the other loosens the nails at each stud.

*If there is no wall sheathing* on a house, the siding can be removed from inside the house, so you don't have to to use a ladder.

*When you have access* from the backside of the siding, you can use a 2×4 to bump out the siding so you have enough room to pry from the other side.

*You can use* a reciprocating saw to remove siding, but it will leave the nail head in the siding and the nail shank in the studs.

back. But be aware that this will leave the heads of the nails in the siding and the nail shank in the studs, a potential safety hazard if someone later cuts into the remaining metal with a saw.

Ideally, the best way to remove siding is to avoiding prying it and simply pull out the nails. However, how do you grab the nail on the front face without damaging the face of the siding? One way is to bump out the siding about ½ in., as described earlier, and then come around to the front of the siding and tap the siding back against the stud (see the drawing on p. 220). This will usually pop the nail out from the siding and allow you to grab the head with a flat prybar. This method works well for stained siding. However, if there are many coats of paint and the wood is brittle (which is the case with redwood), the paint film can be strong enough to cause a wood chunk to tear out with the nail; the hole will need to be filled before reusing the siding.

If your house has 1× wall sheathing, remove it in the same way as the roof sheathing. In some cases, it will have a tongue-and-groove (T&G) or lap profile, and you will have to remove it using the same process as removing flooring to prevent breakage.

(see the drawing on p. 220)

## HERE'S A TIP

### Removing Siding

How you use your prybar can make it easier to remove siding. You can gain a little more leverage if you place the bar as shown in the top photo below—that is, angle the prybar and use the stud as a fulcrum. Placing the bar as shown in the bottom photo below makes it more difficult to get a grip on the board, and you risk damaging the siding.

Tapping siding from the back side pops the nails up, making removal of nails easier from the front side.

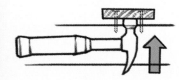

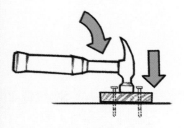

## Removing Walls

The next step in the deconstruction process of this house was to work on bringing down the walls. The best way to do this is in the reverse order that a carpenter would have put them up. In platform construction, walls are typically constructed lying flat on the floor and tilted up and nailed to the floor through the bottom plate and to adjoining walls.

It's best to remove interior walls before exterior ones. Think about which ones come down first, as you want to do it in a sequence to progressively open up floor space. Obviously, you want enough clear space in front of where the wall will fall to accommodate its height and to allow it to come down flat on the floor. Drop the exterior walls (typically inward) last.

There are times when you may not want to bring the whole length of a wall down at once. In this case, you can drop the wall in two or more sections, as shown in "Removing a Long Wall in Sections" on p. 222. Always use a rope to make sure a gust of wind doesn't take the wall over the edge, and be sure to brace the portion of the remaining wall.

### Removing Vinyl or Metal Siding

As with wood siding, to remove vinyl siding you have to carefully pry out the nails, otherwise you will rip the vinyl. It's important to leave the slotted nail holes intact because if the siding is reinstalled it has to be renailed in those slots so the siding can expand and contract. Remove any corner or trim pieces before the siding itself. Metal siding also requires care to remove, stack, and store because it dents and bends easily.

The first step in dropping a wall is to cut it free from adjoining walls.

You may need to cut the nails holding the wall to the floor using a reciprocating saw. Often, the wall will come down without removing these nails, saving you a step.

Once the wall is cut free, push it over using a long 2×4 to stay clear of danger.

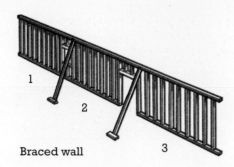

1

2

Braced wall

3

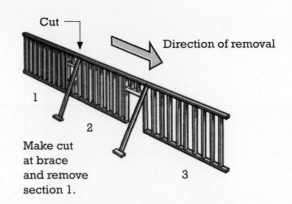

Cut

Direction of removal

1

2

3

Make cut
at brace
and remove
section 1.

Cut

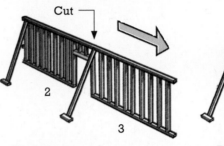

2

3

Cut at brace and
remove section 2.

3

Remove final
section 3.

## Balloon Framing

To remove exterior walls on
the second floor of a balloon-
framed house (where the
outside wall studs are contin-
uous from foundation to
eaves), you will either have
to cut the outside wall studs
at floor level to drop the wall
or remove the top plate, blocking, and window framing and
leave the tall studs standing
until you get to the first floor.
The second option assumes
you want to keep the long
studs intact.

*To remove* an outside wall on the
second floor of a balloon-framed
house, you have to cut the wall studs
before you can drop the wall.

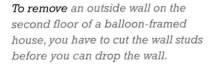

**Skip the Sill Plate**

In houses with slab-on-grade foundations, you'll likely find sill plates of walls that are bolted down to the foundation. It's probably not worth trying to salvage a bottom plate bolted to a floor because of all the bolt holes. Trying to unthread rusted nuts or cutting them off with a reciprocating saw can be frustratingly slow and take more time than it is worth.

**Dropping Gable Ends**

The gable end of the military warehouse shown at right was cut from the sidewalls and pulled down to be disassembled. It was dropped outward because the ground was soft and flat and it was felt that more damage would be done if it were dropped on the building's concrete slab. Also, because the peak was taller than the standing sidewalls, it could have gotten hung up on the way down if dropped inward.

The crew used a series of ropes and poles to help get the wall moving in the right direction. If you try this, make sure everyone is ready to clear the fall area (forward of the wall *and* in back of the wall, in case it goes the wrong way). The basic rule of thumb when pulling anything over is to have $1\frac{1}{2}$ times the height of clear space in the direction of the fall and a clear space at a 45-degree angle from the base. So for a 10-ft.-high wall, you would want to have at least 15 ft. of unobstructed space in the direction of fall.

After you have the wall down, a sledgehammer works well to beat off the bottom and top plates. It's pretty quick and easy to take the wall apart on the ground with the sledge, your hammer, and some prybars. The stairway in the house was also removed, and the procedure was not unlike removing a wall. The stair had to be cut away from its supports with a reciprocating saw and dropped to the floor for removal. The treads and risers were not removed. Though heavy, it would be easy for someone to use this stair for a new application.

## Dealing with nailed posts, beams, and headers

As you take down the house you'll find built-up headers over windows and doors and heavier built-up beams and posts that need to be pried apart for salvage. In construction today, you will commonly find single-member glulam, laminated veneer lumber, I-joist, and composite lumber headers and beams; however, in most old homes 2× lumber was often built up to make larger members.

*Beat off the bottom plate* of the dropped wall with a sledgehammer. Although you want to take care not to strike the lumber too much with such a heavy tool, it can make short work of loosening the top and bottom plates from the wall studs.

*At the end* of a workday, the walls were braced to prevent collapse by wind or inquisitive neighborhood children.

*The stairway* was removed by cutting the supports and dropping it to the floor, keeping the unit intact. It will be easy for someone to integrate this stairway into another building.

*This large built-up frame* was cut loose from the adjoining walls and dropped to the floor. The individual pieces of lumber in the frame were pried apart for easier handling.

If two nailed-together pieces are small, a pair of small prybars works to force them apart (or a prybar and a crowbar, as shown above). The key is to use a push–pull action, which effectively doubles the prying force. Start at one end and work your way down the piece until the boards are separated.

For more leverage, use a larger pair of bars.

On heavier components, you may need two people to pry the boards apart.

Prying apart the individual pieces of a built-up component can be a challenge. Depending on the size of the pieces, you can use a pair of prybars or crowbars to make the job easier. The key is to use a prybar in each hand and pull one against the other to double your prying force. On larger members, you may need two people working together.

## Removing Subfloors

Subfloors support the finish wood flooring or carpet of a house. In older homes, 1× solid wood was used, but it was largely replaced by plywood after World War II. In today's construction OSB has largely replaced plywood. If the subfloor is tongue and groove, the procedure for removal is the same as for finished flooring (see pp. 180-185). Work from the tongue side and move across the floor.

Plywood or OSB subfloors can be difficult to remove without destroying the sheets. If you have access from below, you might be able to pop up the sheets of plywood with the head of a sledgehammer to loosen the nails. However, in the last 30 years or so, construction adhesives have been used between the floor joists and subfloor to prevent floor squeaks, making salvage of plywood or OSB more difficult.

In this house, the subfloor was a 1× Douglas fir board that was of quite high quality and easy to remove. Using taller prybars to remove the subfloor was easier on the back and knees than using a short, flat prybar. If prying does too much damage to the subfloor, you can try using a reciprocating saw to cut the nails (though you will end up with a joist that is full of nail shanks).

*Taller prybars* were used to remove the 1× subfloor.

*Cutting the nails* minimizes wood splitting but leaves the nail shank in the joist, which is a potential safety hazard if it's hit with a circular saw when reused.

## Floor joists

Once the subfloor is removed, the floor joists can be harvested. Because of the age of this house, it was constructed with balloon-framing techniques and had no rim joist, which would have to be removed in newer construction. The cross-bracing between each joist had to be knocked out and then the joist could be laid over and carried off. A sledgehammer worked well for removing the bracing. The stem walls and interior post-and-beam supports on which the joists rested were removed after all the joists were taken out.

## Foundation and final site prep

After the cripple walls (that is, the crawl-space walls) and interior post-and-beam foundation supports were removed, the site was cleaned of all building debris. A skid-steer tractor was used to remove the concrete piers, pads, and other rubble to leave the site ready for the construction of a new house.

*As this house was built with balloon-framing techniques, there was no rim joist to remove from the ends of the floor joists.*

### RIM JOISTS

In today's platform-frame construction, a rim joist is used to stabilize and support the ends of the joists. The joists are hung from the rim joist with joist hangers if not supported on a sill plate or beam for bearing. Before the advent of joist hangers, the rim joist was end-nailed to each joist.

*Removing the floor* bracing with a sledge-hammer frees up the floor the joists, which can then be rolled over and moved to the denailing station.

*A skid-steer tractor* was used to remove the interior concrete piers and to clean up other rubble.

*Only the exterior wall* foundation remained after the unbuilding project; the contractor who was building a new house on the site would remove this wall.

*Working across the floor,* the workers removed all the joists until only the exterior stem walls and interior post-and-beam joist supports remained.

## TAKING DOWN A CHIMNEY

Though this house did not have a chimney, many homes do. If there is one in the house you are deconstructing, you should first assess whether the chimney is safe to work around. If it is leaning or has large cracks, you might want to call a chimney professional for advice. If the mortar is soft enough, you can chisel off bricks one at a time and drop them down the flue. Of course you will have to have a scaffold to do this. If you remove the damper first, the bricks will spill out the fireplace hearth for pickup below. This way the chimney is used as a chute to contain the bricks.

The easier way to remove a chimney is to topple it, which will require some equipment and room for it to fall. It's recommended that you have clearance at least $1\frac{1}{2}$ times the height of the chimney at an angle of 45 degrees to either side of the intended line of fall.

*You'll need some room to topple a chimney with equipment.*

## Denailing

Removing nails and other hardware from lumber is a necessary, but often tedious, part of deconstruction. As we described earlier, you should set up a denailing station that is in as direct a path as possible between where the lumber is being removed and where it will be stacked for banding and loading. Also, locate the station to minimize the distance you have to carry lumber with nails in it.

It's good to keep in mind that you will exert less effort and do less damage to the wood if you pull the nail out in as straight a line as possible—ideally, you want to drive or pull out the nail at the same angle it was put in.

You will always do less damage if you drive the nail out from the point. If you have to pull nails from the head, you need to be able to get the tool underneath the nailhead to pull. The cat's paw is effective, but inevitably you will do some damage to the wood to get the grip needed (see the photos on p. 232).

*The tried-and-true method* of removing a nail is to pound out the nail from the point with a hammer and once you have the head exposed enough grab with the claw or with a prybar to pull it out. If you need more leverage, use a wrecking bar.

**HERE'S A TIP**

## A Little Nail Leverage
You can put some physics to work when pulling nails by raising the fulcrum of the prying tool with a block of wood or another tool. This not only gives you mechanical advantage but also all allows you to pull the nail out of the hole in a straighter line, reducing friction and the force you need to pull.

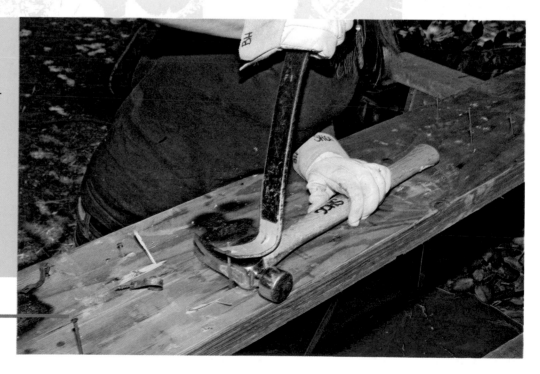

*You can gain extra leverage* and use less effort if you set a block of wood or a hammer under your prying tool.

Life has gotten easier for nail pullers since Jon Giltner, of ReConnx, Inc., started importing and selling a pneumatic denailer. Marketed as the Nail Kicker, it was originally developed in Japan, though Giltner has modified and redesigned the tool to American standards. The Nail Kicker works from the point side of the nail to drive the nail out of the wood. This tool will save you a lot of time in denailing; some users say it's four times faster than hand denailing. Specific claims aside, it is certainly much faster than using a hammer and prybar. If you do a lot of denailing, this is the tool to have. Because it is pneumatic, you will have to have a compressor on site. Also, to keep nails from flying all over, it's important to either shoot the nails into a padded trash barrel or onto a padded surface, such as old carpet or carpet padding. Hanging a tarp or carpet from the sawhorses can help protect the operator.

*To use a cat's paw,* *pound in the claw with your hammer, rotating the tool so it drives under the nailhead as you pound. Once you have enough bite under the nail, lever the tool to remove the nail. This tool is really only good on framing lumber, because of the damage it does to the wood.*

*A useful feature* of the Nail Kicker is that you can use the barrel of the tool to straighten a bent nail before shooting. This helps minimize damage to the board and, more important, allows you to direct the nail into a barrel or carpet pad. If you try to shoot the nail in a crooked position, it may not eject, or if it does, it could shoot off in any direction, often toward you (which is why you should hang a tarp for protection from errant nails).

HERE'S A TIP

**Nail Cleanup**
Cleaning up nails regularly keeps the site safer and truck tires happy. A magnetic sweeper is a great tool for this. It's good to keep a 5-gal. plastic pail handy to collect the nails for recycling.

*A magnetic sweeper* can make nail cleanup easy. It's easy to remove the nails from the magnet by pulling the handle.

**The Nail Kicker** is powerful enough to remove nails from engineered lumber, such as the engineered wood beam shown here.

It's always a challenge on the job site to minimize bottlenecks and keep material flowing. The denailing station is often such a bottleneck. With only one Nail Kicker available on this project, there were times when the lumber couldn't be denailed fast enough. Material coming off a building seems to surge at certain stages—for example, when rafters or joists are removed. When possible, keep denailers and lumber bundlers busy between these surges cleaning up any backlog of material so they are ready for the next wave.

## Stacking and Loading

After denailing, all the lumber was stacked into bundles of like type (framing, siding, flooring) and width (2×4, 2×6, 2×8, and so on). Try not to mix types or widths in the same bundle, as you will only have to sort it later. It's always better to make bundles of like length if possible because it's easier to maximize the volume you can put on the truck. Also, the individual pieces of lumber in a single-length bundle support each other. If you have various lengths in the same bundle, the longer pieces will have no support and will flop around and get damaged. If you do have to stack mixed lengths, make sure one end of the stack is square. It makes it easier to tally the lumber.

*When possible, try to make bundles of lumber, siding, and flooring that are the same length.*

The stacked and banded bundles of lumber, flooring, and siding were loaded onto a flatbed truck using a skid-steer tractor. Although a forklift would have been preferable, the contractor on this job had a skid-steer with forks. Because all lumber bundles had been properly sized for the capacity of the tractor, there was no problem lifting each bundle onto the truck. Also, many bundles could be loaded on one truck because each was sized at half the bed width. Care was taken to use lumber bolsters (see p. 99) to make unloading easy with a forklift at the other end of the trip.

*For stacks with different lengths of lumber, make sure one end of the stack is square to make it easier to tally the lumber.*

**HERE'S A TIP**

### Keeping the Lumber Together

You'll need to band the lumber to keep it together for loading, unloading, and transport on a truck. If you deal with a lot of lumber, it's worth investing in a set of banding tools.

*On this project, a skid-steer tractor was used to move banded bundles of lumber for shipment.*

***Try to pack*** *the truck to maximize the material you can get in one trip (without exceeding the weight limit of the truck, of course). If you stack lumber in bundles no wider than half the truck bed, you can set two bundles side by side.*

*When stacking* the bundles, make sure to use bolsters between them.

*Never ship a load* of lumber without strapping it down securely. Sturdy trucking straps with a tensioning ratchet are required.

The materials salvaged from this this project were delivered to resale warehouse of The ReUse People in Oakland, California, the deconstruction managers on the project. The quantity of material collected was impressive for this modest-size home. About 7,000 bd. ft. of framing lumber was salvaged, along with 26 4×8 sheets of plywood, 17 half-sheets of plywood, 18 doors, 8 windows, 19 cabinets, a fireplace insert, and a fireplace mantel. Dozens of other items, including lighting fixtures, ceiling fans, a shower stall, whirlpool tub, water heater, staircase, and porch railing and posts were also collected. In addition, we salvaged almost 5,000 lin. ft. of Douglas fir flooring and 2,300 ft. of redwood siding.

In the end, this deconstruction project was worthwhile, and the salvaged materials quickly sold at The ReUse People's warehouse. The owner of the house was also happy because the building was removed less expensively than if it had been demolished. Finally, the materials were diverted from the landfill and instead were reused—which is always the goal of any deconstruction project.

Since you've read this far, we hope you have found the book useful. Whether you are a homeowner, a building contractor, or someone seriously thinking of starting a deconstruction business, you *and* our planet will benefit from the salvage and reuse of building materials.

Although no one book can show you everything, there is enough information in these pages to serve as a reference as you undertake your own project or as a resource for teaching newcomers to the field of unbuilding. In either case, by deconstructing a building you'll be doing something positive for the environment. You'll expend some sweat; but in the end, you'll have helped extend the life of our nation's resources and have kept quality materials out of the landfill.

*Barry Stup of The Woods Company started his reclaimed flooring business by salvaging barns destined for the bulldozer.*

"My philosophy is simple: Save stuff, don't waste, reuse."

# Saving Barns for Reuse

## Barry Stup

Barry Stup of the Woods Company began dismantling barns in rural Pennsylvania in the 1970s with the intent to use the materials in new construction and renovations. "Early in my career, I concentrated on interior woodworking projects such as staircases, lofts, and trim work. By the mid-1980s I established The Woods Company as a flooring manufacturer, recognizing the greater market for antique wood flooring." Twenty-five years later, Barry's company has grown into one of the largest reclaimed lumber manufacturers, with over 120,000 sq. ft. of millwork space and 40 employees.

Barry was greatly influenced by the environmental movement in the 1970s, which made him aware of the need for recycling and the logic behind it. He became aware of the large number of barns that were neglected and decaying; these were most often burned or land-filled when the property was being redeveloped. "One of my earliest realizations was that some of the barns in Maryland and Pennsylvania were occasionally built out of chestnut, which is one of the most rare and beautiful of the native hardwoods," says Barry. "The timbers were often in good condition and instead of being salvaged were simply wasted."

In the past 10 years Barry's business has become much more mainstream, spawning an array of competitors and more pressure to remove old barns. Due to increasing development, many barns have become available for deconstruction. Because of modernized farming techniques and changing land use, many old barns have been rendered obsolete. They have often outlived their usefulness on farms and are being replaced by more modern structures. Rising maintenance costs lead these barns to be neglected by their owners leading to their eventual decay and demise.

"In a way, barns have become a victim of their own popularity; instead of being seen as valuable structures they are seen as valuable pieces and parts. It may often be unnecessary to dismantle a barn that is in good condition, but it may also be more profitable to do so. To save good structures, I would like to see more emphasis on salvaging the timber frames for reuse. If you can save the frame, you can effectively save the barn. It can be rebuilt elsewhere, thereby regenerating the vernacular architecture."

We live in a wealthy society that is used to wasting resources. However, in less affluent cultures, buildings and materials are virtually always recycled in some form or another. It makes environmental and economic sense. "The idea of giving a material a new life may sound like a cliché," admits Barry, "but in fact it is highly sensible and fundamentally sound. By recognizing the inherent value in a material, new products can be created from something that might otherwise be considered waste. My philosophy is simple: Save stuff, don't waste, reuse."

# Estimating the Weight of Building Materials

The following weights of materials are provided to help you estimate the total tonnage of materials you need to deal with in your unbuilding project. These numbers will help you estimate how much weight you will either have to move for reuse or transport to the landfill.

## Materials Measured by Area

| Material | Weight per sq. ft. (in lb.) |
|---|---|
| **Roofing (per layer)** | |
| Wood shingles | 3.0 |
| Asphalt shingles | 2.5 |
| Builtup (3-layer flat roof) | 6.5 |
| Wood shakes | 6.0 |
| Clay tile | 9.0–14.0 |
| Slate (¼") | 10.0 |
| **Repetitive Framing (wall, floor, rafter)\*** | |
| 2x4 | |
| 12 in. o.c.\*\* | 1.4 |
| 16 in. o.c. | 1.1 |
| 24 in. o.c. | 0.7 |
| 2x6 | |
| 12 in. o.c. | 2.0 |
| 16 in. o.c. | 1.7 |
| 24 in. o.c. | 1.0 |
| 2x8 | |
| 12 in. o.c. | 2.6 |
| 16 in. o.c. | 2.2 |
| 24 in. o.c. | 1.3 |
| 2x10 | |
| 12 in. o.c. | 3.4 |
| 16 in. o.c. | 2.8 |
| 24 in. o.c. | 1.7 |
| 2x12 | |
| 12 in. o.c. | 4.1 |
| 16 in. o.c. | 3.2 |
| 24 in. o.c. | 2.1 |

\* most softwoods (35 lb./ft$^3$)

\*\*o.c. = on center

| **Sheathing** | |
|---|---|
| ½-in. plywood | 1.5 |
| ¾-in. plywood | 2.3 |

| Material, continued | Weight per sq. ft. (in lb.), continued |
|---|---|
| ½-in. OSB | 1.6 |
| ¾-in. OSB | 2.5 |
| 1⅛-in. plywood | 3.4 |
| 1-in. (nom.) board sheathing | 2.9 |
| ½-in. gypsum board | 2.2 |
| ⅝-in. gypsum board | 2.8 |
| Lath & plaster (1" thick) | 8.0 |
| **Flooring** | |
| Hardwood flooring (solid oak) | 4.0 |
| 2" (nom.) decking | 4.3 |
| Linoleum | 1.5 |
| ¾" ceramic tile or quarry tile | 10.0 |
| **Other Materials** | |
| Glass pane (⅛") | 1.7 |
| Acoustical tile | 1.0 |

## Materials Measured by Volume

| Material | Weight per cu. ft. (in lb.) |
|---|---|
| Gravel | 90 |
| Sand (dry) | 100 |
| Sand (wet) | 130 |
| Concrete | 150 |
| Brick | 150 |
| Dirt | 80–110 |
| Garbage (average) | 30 |
| Marble | 160 |

## Lumber Weights*

| Size | Weight per lin. ft. (in lb.) |
|---|---|
| 2×4 | 1.4 |
| 2×6 | 2.0 |
| 2×8 | 2.6 |
| 2×10 | 3.4 |
| 2×12 | 4.1 |
| 3×6 | 3.3 |
| 4×4 | 3.0 |
| 4×6 | 4.7 |
| 4×8 | 6.2 |
| 4×10 | 7.9 |
| 4×12 | 9.6 |

* most softwoods (35 lb./ft.$^3$)

## Nominal and Actual Lumber Dimensions

**Standard 2x Lumber Sizes**

| Nominal size | Actual size (in.) | Nominal board footage (per ft. of length) |
|---|---|---|
| 2×4 | 1½ × 3½ | 0.7 |
| 2×6 | 1½ × 5½ | 1.0 |
| 2×8 | 1½ × 7¼ | 1.3 |
| 2×10 | 1½ × 9¼ | 1.7 |
| 2×12 | 1½ × 11¼ | 2.0 |

# Tools for a Full Deconstruction

In chapter 4, we listed the basic unbuilding tools you'll need to get started (see p. 120). The tools listed below are those we recommend if you're tackling a whole-house deconstruction. (Tools are listed alphabetically, not by preference.)

- ax
- board lifter
- bolt cutter
- branch trimmer
- carpenter's square
- chisel (brick)
- cold packs for sprains
- crowbars (a range of sizes as required)
- dustpan (extra-large)
- fall protection gear
- fire extinguishers (ABC)
- gang box (electrical)
- generator (and gas can)
- half-mask respirator (and extra filter cartridges)
- hand sanitizer
- level
- lifeline (self-retracting)
- magnet for nails
- moisture meter
- Nail Kicker nail remover
- padlock

- pick ax
- plastic buckets (5-gal.)
- plywood sheets
- post-hole digger
- roof anchor (re-usable)
- rope
- scaffolding (rent or buy)
- screwdrivers (Phillips, slot head)
- shovels (roofer's, snow)
- tape ("caution," duct)
- tarps
- tie-down straps
- Tyvek suit (with booties)
- warning signs ("Danger Hard Hat Area")
- water cooler and potable water
- water hose
- wheelbarrow
- work lamp
- wrecker bar (two-clawed)
- wrecker's adze
- wrenches (adjustable open-end, socket)

# Resources

Most of the tools and safety equipment we use for deconstructing a building are widely available and can be easily found in your local hardware store, safety products company, home improvement box store, or online. Some of the specialty tools we've talked about may be a little harder to find. Here's some contact information:

**Fulton Corporation**
303 8th Avenue
Fulton, IL 61252
(815) 589-3211
www.fultoncorp.com
*Wrecker bar*

**Klein Tools, Inc.**
450 Bond Street
P.O. Box 1418
Lincolnshire, IL 60069-1418
(847) 478-0625
www.kleintools.com
*Grizzly Bar*

**Jefferson Tool, LLC**
P.O. Box 31535
Charleston, SC 29417-1535
(843) 556-0455
www.nailextractor.com

**Malco Products, Inc.**
14080 State Highway 55 NW
P.O. Box 400
Annandale, MN 55302-0400
(800) 596-3494
www.malcoproducts.com
*Roofing shovels*

**Metcalfe Roush Forge and Design, Inc.**
18 Waldroup Road
Brasstown, NC 28902
(828) 835-7313
www.metcalferoush.com
*Sheathing/decking lifting tool, flooring pry tool*

**Miller Fall Protection**
Bacou-Dalloz US Headquarters
910 Douglas Pike
Smithfield, RI 02917
(800) 343-3411
www.millerfallprotection.com
*Safety harnesses, roof anchors, and other fall protection*

**ReConnX, Inc.**
P.O. Box 3009
Boulder, CO 80307
(303) 554-8554
www.reconnx.com
*Nail Kicker*

**Vulcan Tool Company**
52 Sharp Street
South Hingham, MA 02043
(800) 247-4770
www.vulcantools.com
*Wrecker's adze*

# Some Useful Websites

While the following list is by no means exhaustive, the websites below contain useful information on materials reuse and building deconstruction. Search for "building deconstruction" or "building materials reuse" to find other websites.

### Alameda County Solid Waste Management Authority

www.stopwaste.org/fsbuild.html
Provides green building manuals and resource guide for residential construction and remodeling, including information on deconstruction and reuse.

### Architectural Salvage Exchange

www.salvageweb.com
Online exchange and forums for architectural salvage.

### Architectural Salvage News

www.architecturalsalvagenews.
  com/pages/1/index.htm
Newsletter for the architectural salvage industry.

### Building Materials Reuse Association (BMRA)

www.buildingreuse.org
National nonprofit educational association representing the deconstruction and building materials reuse industry.

### Building Green

www.buildinggreen.com
Information on green building and deconstruction.

### Build Recycle.net

www.build.recycle.net
National reused materials exchange.

### C&D Recycler

www.cdrecycler.com
Information on the latest products/ services, news, events, classifieds, RFPs, and article archives for the construction and demolition waste-recycling industry.

### California Integrated Waste Management Board Green Building Design and Construction

www.ciwmb.ca.gov/GreenBuilding/
Sustainable building toolkit for project managers, blueprint for state facilities, performance standards, project designs, and case studies; includes a large section on deconstruction and reuse.

### City of Seattle, Sustainable Building

www.cityofseattle.net/
  sustainablebuilding/
Current city projects, city policy info, facility standards, incentive programs, and resources; includes guidance for reusing materials in construction.

### Construction Recycling Guide

www.metrokc.gov/dnrp/swd/
  construction-recycling/index.asp
Offers a contractor's guide to saving money through job-site recycling and waste prevention, calculating cost-effectiveness, and contract design specs.

### Deconstruction Institute

www.deconstructioninstitute.com
Resources for deconstruction and reuse.

### EPA Construction and Demolition

www.epa.gov/epaoswer/non-hw/
  debris-new/index.htm
Defines C&D debris, lists disposal methods and additional resources.

### EPA Green Buildings

www.epa.gov/greenbuilding
Describes and links to green building materials, energy efficiency and renewables, indoor environment and waste.

### Habitat for Humanity ReStores

www.habitat.org/env/restores.aspx
Listings for HfH ReStores in the US.

### Hamer Center for Community Design

www.hamercenter.psu.edu
Community design and deconstruction research and projects.

### Powell Center for Construction and Environment

www.cce.ufl.edu
Research and development information on green building, including deconstruction.

### U.S. Green Building Council

www.usgbc.org
Members work to develop design guidelines, policy positions, educational tools and industry standards. Information on certified projects and accredited professionals.

### USDA Forest Products Laboratory

www.fpl.fs.fed.us
Information on wood usage, wood products, and wood properties. Thousands of publications, many of which deal with deconstruction and the salvage and reuse of lumber and timber.

### WasteCap Wisconsin

www.wastecapwi.org
Information on waste reduction and recycling assistance to businesses.

# Unbuilding Pioneers

### Kevin Brooks

Kevin Brooks Salvage
1320 North 5th Street
Philadelphia, PA 19122
(866) 364-8747
ktbsalvage@verizon.net
www.kevinbrookssalvage.com

### Pete Hendricks

Deconstruction Consultant
1388 Jenkins Road
Wake Forest, NC 27587
(919) 556-2284

### Sally Kamprath/Kathy Burdick

The ReHouse Store
1473 East Main Street
Rochester, NY 14609
(585) 288-3080
info@rehouseny.com
www.rehouseny.com

### Linda Lee Mellish

ReStore
3016 East Thompson St.
Philadelphia, PA 19134
(215) 634-3474
info@re-store-online.com
www.re-store-online.com

### Ted Reiff

The ReUse People of America, Inc.
9235 San Leandro Street
Oakland, CA 94603
(510) 383-1983
Info@TheReusePeople.org
www.thereusepeople.org

### Jim Stowell

The Whole Log Lumber Company
688 Blueberry Farm Road
Zirconia, NC 28790
(828) 697-0357
(866) 912-WOOD
info@wholeloglumber.com
www.wholeloglumber.com

### Barry Stup

The Woods Company
985 Superior Avenue
Chambersburg, PA 17201
(717) 263-6524
(888) 548-7609
woodfloors@thewoodscompany.com
www.thewoodscompany.com

### Max Taubert/Peter Krieger

Duluth Timber Company
P.O. Box 16717
Duluth, MN 55816
(218) 727-2145
info@duluthtimber.com
www.duluthtimber.com

### Jen Voichick

Habitat for Humanity of Dane County
ReStore
208 Cottage Grove Road
Madison, WI 53716
(608) 661-2813
jvoichick@restoredane.org
www.restoredane.org

# Index